CAR HACKS

- 126 tips & tricks to improve your car
- Quick and simple car cleaning tricks
- Use household objects as storage
- Entertain the family on long journeys

Craig Stewart

CONTENTS

Introduction   4

1 INTERIOR HACKS   6

Add 12v sockets throughout car   8
Home-brewed dehumidifier   10
Stocking up on bottles   11
Keep cupholders clean and stop rattles   12
Binder clip phone mount   13
Rubber band phone holder   14
Turn your car's ashtray into a smartphone dock   15
Routing charging cables out of the way   16
Lining door bins to stop rattles   17
Carabiner headrest hooks   19
Bungee cord to stop bags tipping in boot   20
Use your seat warmer to keep food warm   21

2 EXTERIOR HACKS   22

Maximise your car's life   24
Use nail polish to fix paint chips and scrapes   29
Use a plunger to fix dents   30
Use hand sanitiser to de-ice keyholes   31
Rub out scuffs with toothpaste   32
Temporary wiper blade replacements   33
Temporary window damage fix using clear nail polish   34
Plastic bag door mirror ice preventer   35
Use socks to protect wiper blades   36
East is East   37

3 GARAGE HACKS   38

Protecting tyres from overspray   39
Create a handy map of bolts and screws   42
Magnet in breast pocket to keep tools handy   44
Take photos of trim locations   45
Labelling wires when working on car electrics   46
Find water ingress   47
Bungee to hold brake caliper   52
Prevent door damage with pool noodles   53
Hang a tennis ball at the back of your garage roof as a parking guide 54
Use your phone's selfie camera to visualise awkward access points   55

4 TRAVEL HACKS 56

Maximise your car's efficiency (hypermiling) 57
Cupholder expander 62
Photograph car parking location 63
Tablet holder hacks 64
Use a Magna Doodle as a combo car toy and food tray 66
Wipe-off markers for drawing on windows 67
Snack packs 68
Make a glove box kit 69
Avoid road rage 70

5 STORAGE HACKS 74

Add a curry hook 76
Make boot dividers and add boot storage 77
Install a semi-permanent cool box in the boot 79
Roof box hacks 80
Create a backseat organiser with a shoe rack 81
Mesh bungee net can add extra storage 82
Utilising previously useless space 83
Something out of muffin 84
Tiny containers to divide up storage areas 85
Coin tubes 86

6 CLEANING HACKS 88

Use olive oil to polish interior trim 90
Use a blade to clean your windscreen 91
Coffee filter dusters 92
DIY air fresheners 93
Cat litter moisture removal 94
Use plastic breakfast cereal boxes as waste bins 95
Make your own grit guard bucket 96
Use toothpaste to make headlights minty fresh 98
Baking soda for smelly upholstery 99
Screwdriver cloth to replace Q Tips 100

7 MODERNISING HACKS 102

Dashboard trim respray or hydrographic 104
Mirror/exterior trim wrap 106
Window tinting 107
Headlight updates 108
In-car Wi-Fi 110
Get in-car performance data readouts 111
Smartphone dashcam 112
Add aux input to stereo 113
Add USB sockets in rear 114
OEM+ upgrades 115
Add interior LEDs 118

INTRODUCTION

What is car hacking? Car hacks are small changes you can make to your car and its environment to improve your motoring life. From minor modifications to using clever solutions to upgrade your vehicle, car hacks allow everyone – even those with no mechanical knowledge – to upskill their vehicle with minimal time, effort and money. Get the latest gadgets in your car without shelling out on new ones, make long journeys less painful and ensure the kids are entertained – we'll show you how.

Driving, owning, repairing and maintaining a car can be expensive, frustrating and time-consuming. This book is not, however, about corner-cutting, nor does car hacking encroach into areas that could endanger the car hacker either while working on a vehicle or when driving. *Car Hacks* is here to explain how to use the things you have around your home to improve your car life, and improve your well-being in the process. From ensuring you never lose a screw when repairing your car, to spending less on fuel, and using cereal boxes to stop cleaning your car so often, this book will open your eyes to the joys of car hacking.

CHAPTER 1

INTERIOR HACKS

On average, we spend more time in our cars than we do socialising. However, car interiors are designed to appeal to a broad section of the public, and that means they're not tailored for your specific needs. Therefore, much of the time you spend in your car will be riddled with frustrations. This chapter outlines some simple hacks to create the car interior that you need, from adding power sockets and storage where you need them to eradicating annoying rattles.

Add 12v sockets throughout car

A great aspect of the explosion in popularity of SUVs, crossovers and other vehicles aimed at an active lifestyle is that they frequently come with useful 12v sockets throughout the interior. Even better, this feature has also trickled down to everyday family cars. But if you own an older or more basic car, you can grab a few tips from the world of camper vans to add more 12v sockets in the centre console, rear seat area or even in the boot, to power a cool box or even a shower.

How you do this will vary in detail depending on which car you own. So this hack will take you through the basic process of adding a second 12v outlet to the dashboard, but the process is the same for locating a socket in various areas around the interior of your car.

1. Get hold of an aftermarket 12v socket

These are also known as cigarette lighters, of course. You can get them from auto parts stores, eBay and more. Some have a built-in light, some don't – choose whichever suits your application best.

2. Decide where to place it

Find a suitable position, ensuring that there's space behind for the socket's internals. If you can remove the panel easily, it's a good idea to do that before drilling. Mask up that area with tape and draw a circle on the masking tape that is slightly smaller than the diameter of the 12v socket.

3. Drill a hole

First drill a pilot hole in the centre of the drawn circle. Then use that to centre the correct-size drill bit and drill the hole. If you don't have a big enough drill bit, use the biggest you have and file out the rest of the hole.

4. Insert the socket

Remove the masking tape and insert the socket into the hole. Then, from behind, slot the casing or threaded ring on to the socket body (this will depend on the fitting of the socket you have purchased).

5. The electrics

Now disconnect the negative terminal of the car battery. Run a wire containing a 35 amp inline fuse from any permanent live feed to the centre terminal on the socket. Join a wire to the earth terminal on the casing and run it to a nearby metal point on the body. Then reconnect the battery and test the socket.

For more in-depth information about this, particularly if you intend to add a leisure battery or 240v, I thoroughly recommend Haynes' *Motorcaravan Manual* 3rd Edition (book number H5124), which has a full section on this in Chapter 5.

Home-brewed dehumidifier

Use a screwdriver as a funnel

Insert the end of a longish screwdriver into the top of your oil filler, hold it near vertically, then simply pour your oil down the shaft of the tool and you won't spill a drop.

Cars get damp and musty the longer you own them, regardless of how efficient you are with the heating and aircon. Plus that new-car smell begins to disappear the moment you drive out of the dealership, and is slowly replaced by your body odours, fast-food remnants, old coffee and whatever you've brought in on your shoes.

A great hack to dehumidify your car while giving a fresh aroma to the interior is to use a few herbal teabags. Strategically place them around the car when you leave it overnight – one each in the door pockets and one on the top of the dashboard near the windscreen – and your vehicle will be fresh as a daisy in the morning.

Don't be tempted to pop them in a mug for your morning cuppa, though. Chuck them in the compost or the food recycling bin.

A dryer, better-smelling interior will make you feel better about your car every time you get in to drive, aiding relaxation and well-being – all this from a few teabags!

Stocking up on bottles

Irritating noises can grind away at your patience while you're driving, and anything that does that is likely to affect your mood – and the natural progression from there is being a less considerate driver. The problem is that car storage areas for cups, bottles and other essentials while driving are not specifically designed for the products you own. That brilliant reusable plastic water bottle or aluminium flask can become a major irritant in your car's cupholders if they are slender and rattle around while the car is on the road.

The solution to this can be found in your sock drawer. Grab an old sock, or cut off the bottom of a pair of thick tights, and pull it over the bottom of your bottle or flask. Now it should fit into your bottle storage area without that annoying noise – and if not try folding the sock back on itself until you get a snug fit. This will probably be required if you're using the bottle holder that you find in some cars' door pockets, as they can be quite capacious.

 ## Keep cupholders clean and stop rattles

Cupholders are super-useful, there's no doubt. They first appeared in the form we now know them in GM minivans in the 1980s, and then became a firm fixture in family cars during the 1990s. The US cupholder phenomenon shows no sign of letting up, with new vehicles being produced every year that feature more and more of the recessed beverage tubs.

They're not perfect, though. They fill with belongings, dust and crumbs, and in older or cheaper cars they will be fashioned of hard plastic, meaning irritating rattles. Scooping out cupholder detritus can be onerous in cars that don't feature removable inserts, but fortunately there is a little hack that can improve things immeasurably.

Head to the kitchen and grab a couple of silicone cupcake cases and drop them in your car's cupholders. The design of these cases means they should fit snugly in all cupholders, and now you can easily whip out your new liners to empty out any muck. Plus loose items won't rattle around, banging against hard plastic while you're driving: instead they will be bouncing off soft silicone.

Binder clip phone mount

Satellite navigation units are so over – we all carry better, free, constantly updated directional maps in our pockets on our smartphones. Satnav devices do have one major plus point, though – they come bundled with holders that attach to the windscreen, dash top or centre console. Your phone doesn't have that, although there are various aftermarket solutions. But why spend money on a hunk of plastic that may well require replacing every time you upgrade your phone when you can hack a perfectly serviceable smartphone mount? Why indeed.

There are numerous ways to use ingenuity to safely mount your phone in your car, using different household items or extremely cheap products each time. This is my favourite, though, mainly because it involves two great hacking tools: a binder clip (or fold-back clip) and an elastic band. And that's it! Here's how to do it.

1. Bend the binder clip
Binder clips are wonderful things, but for this application you will need to modify them a bit. Pinch the wire parts of the clips and remove them. Now bend the curved ends of each wire piece by about 35°–45° using two sets of pliers. Once that's done, put the clips back together with the newly bent bits pointing inwards.

2. Add rubber band
This is the smart bit. Get an elastic band and stretch it around the wire ends near the clip end. Keep doubling the band around until it's taut but not too tight. Test fit your phone to ensure that it's able to be clasped in the wire ends – if it's too tight, unwind one circuit of the elastic band; if it's not tight enough then wrap the band around once more.

3. Attach to car vent
Now attach the binder clip to the central air vent nearest the driver's side of the car. Once attached, secure the phone into the wire area of the mount by stretching them apart and clipping it in. To avoid any damage to the phone from the wire parts during lengthy use of the mount, it's probably best that it's in a case.

Rubber band phone holder

There are other methods of securing your phone that take even less DIYing. You can use just the elastic band from the binder clip hack. Yes, really. It's not a long-term solution in the way the binder clip hack is, but it will get you by on a short journey.

Get hold of a standard rubber band and loop it through the air vent in the dashboard. From there you should be able to stretch the band around your phone and it should stay fairly steady and in your line of sight – perfect for giving you directions on the phone's navigation app in an emergency.

Turn your car's ashtray into a smartphone dock

Elsewhere in this book I have explained how to create a phone holder for your car with minimal effort and outlay. But come closer and I'll tell you a secret: you may have a phone holder in your car that costs absolutely nothing.

Virtually every car over a certain age came with an ashtray in the dashboard as standard. They are less common nowadays unless you spec a 'Smoker's Pack' in the options list, but there are still plenty around. Now this trick won't work for every car but it will in many. Simply pull out the slide-out ashtray and there should be an area to 'dock' your smartphone revealed to you. If this is too unstable or your phone slides off too easily, a couple of blobs of Blu-Tack should provide enough resistance to stop the device from sliding into your lap.

OK, this isn't going to retain your phone during long trips or spirited driving, and it's likely to be quite far from your line of sight to be genuinely useful as a navigation tool, but it may help get you home at a push, providing you with the ability to glance briefly at the screen, and it will certainly provide controls for your music within easy reach.

Routing charging cables out of the way

Once you have a phone or tablet securely fastened where it's most useful to you, you have another problem to deal with. Namely, cables. Smartphones are amazing devices that can provide you with the tools to do just about anything on the move, but one area where they lag behind their 'dumb' phone predecessors is battery life. Therefore you need the ability to charge them in your car.

While 12v sockets and even USB sockets are not difficult to find in modern vehicles, there's still a problem – they are placed in a position that's entirely likely to leave you wrestling with a dangling cable. This intrusion into your driving space is at the very least an annoyance but also potentially dangerous if it interferes with your ability to get to the gearstick or handbrake in a hurry.

The answer is to route cables along the existing interior architecture, and the requirement is that the cable clips should not be permanently attached – after all, that would look ugly and may not suit every requirement. The last thing you want to do is stick 50 adhesive clips around the centre console to deal with every potential cable route.

The car hacking tool of choice here is the ever-useful binder clip. Small binder clips, sized around 20mm, will do the job perfectly. Poke the end of the charger cable through the widest gap in the wire part of the clips, and thread four or five clips on to the cable. Then attach the clips to any parts of the centre console and dashboard that you can in order to route the cables out of harm's way – vent grilles, console edges, ashtray lip ... anything your car has that can be clipped to will do.

Once you've finished with the device, leave the clips on the cable and store it in the glovebox for next time. Having a car-specific charging cable will save more time and effort.

Lining door bins to stop rattles

There are so many useful storage areas in cars – not just the family-oriented school-run favourites like modern people carriers, minivans and SUVs, but cars through the ages. Glovebox, door pockets, armrest boxes, cassette/CD storage, various cubby-holes… In most cars, however, many of these seemingly handy nooks and crannies are woefully short of thought from the manufacturer.

Hard plastic is the order of the day as the car-makers keep things down to a cost, and that means your belongings rattle and bang as the car lurches left and right, and smaller objects become a melange of dust, dirt, fluff, long-lost trinkets and half-sucked sweets.

Car hackers can do better than this, particularly in the door bins. These longish, deepish storage areas could be useful were it not for the usually dark, usually hard form they come in. There are numerous hacks to solve this. First, get hold of that dimpled silicone mat that's used to line cupboards and drawers, and cut it to size to fit in the bottom – instant premium door furniture that provides a soft and forgiving storage zone.

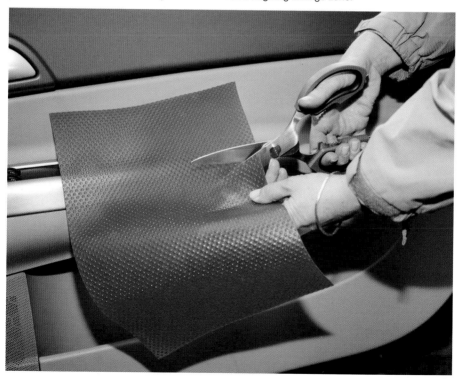

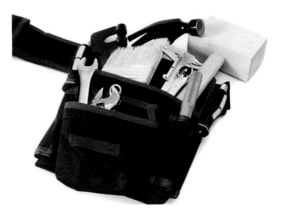

Another plan: if the size of your door pocket allows, buy some drawer dividers – these come in cloth form for dividing up sock drawers from the likes of IKEA, or there are silicone-type dividers for sectioning stationery drawers and craft boxes. These can be picked up for a few quid on auction websites and used to make your door pockets much more car-life friendly. They can be removed individually for cleaning, and will ensure that your trinkets and travel sweets don't become entangled.

One final trick here, particularly if you need to transport a large number of small items such as pens, notepads, compact cameras, torches, sunglasses and so on, is to get hold of a tool pouch that fits snugly in your door pocket. The variety of pockets and holders in these pouches is ideal for any number of items you may require when out and about.

You can even use hook-and-loop adhesive fasteners to make a more permanent attachment to the door panels with the pouch's top sleeves.

TOOL HACK #2

Use a pencil as lubrication

The graphite in pencils is great at providing a slippery surface if you need one quickly, as you always keep a pencil handy. Just rub it on bolts, pins or other parts to help lubricate them.

Carabiner headrest hooks

Another great hacker's tool is the carabiner, or D-ring. Something that can be clipped to anything else to become an instant hook is always going to be an essential element of the car hacker's arsenal, and that's effectively what the carabiner brings to the party.

In this hack, the carabiner aims to end the agony of spilt shopping and tipped takeaways. Clip a carabiner to one of the front seat headrest supports. Now you can hang a carrier bag from the back of the chair, rather than dumping it in the footwell or in the vast cavern of an empty car boot.

Now when your car hits the twisties after a trip to pick up the Friday night carry-out or grab a grocery top-up, the chow mein and cans of beans won't have to be scooped up from the floor mats and under the seats. Thanks, carabiner!

Bungee cord to stop bags tipping in boot

The humble carabiner can save your floor from becoming home to the contents of any carrier bags you decided to place there. Now that's all fine and dandy if you only have a couple of bags to deal with, but what if you have several bags of groceries and a car full of people?

We've all been there – carefully placing the shopping bags in the boot of the car, wedging them against each other as best we can. Then at some point during the short trip home from the supermarket, just one circuit around a mini-roundabout will summon the magic car-boot elves who proceed to pull the breakables, perishables and squashables out of your bags and play football with them in the confines of your car's boot space.

And if your car has a hatchback, look out when you pop it open – because you just know that the glass bottles will be waiting to roll out on to the ground; the final indignity.

This is the perfect car hack to ensure the big shop remains in its shopping bags and not decorating the boot. Another great friend of the car hacker is the bungee cord, and it comes into its own here. Attach one end of an appropriately sized bungee cord to any anchor point on one side of the boot interior (you may need to add your own anchor points – a pair of eye bolts would be a good permanent addition in the right area). Then thread the bungee cord through the handles of the shopping bags before attaching the other end of the cord to an anchor point on the other side of the boot.

Another alternative is to combine with carabiners and have a bungee cord ready-threaded with carabiners attached to the interior of the boot, awaiting your shopping bags. Clip the handles on to the carabiners and marvel at the miraculous way that your groceries remain in the bags at the end of the journey.

Use your seat warmer to keep food warm

Car hacking isn't just for budget motoring. If you're fortunate enough to own a car with heated front seats, we have just the car hack for you. When you've been sent out to pick up the spoils on pizza night (and they always have better offers if you collect, right?), here is the perfect hack to ensure your food is piping hot when it lands on the dinner table.

Put on the heated passenger seat when you head off for collection, and it'll be nice and warm when you arrive at your pizza takeaway of choice. Once you've collected your meal, place the pizza box on the passenger seat and voilà – it'll still be toasty on your arrival home.

Have you also bought sides, such as potato wedges and garlic bread? Bring a cool box or cool bag with you and place them in there for the journey home. Cool boxes and bags are insulated, which means that they not only keep cold things cool but also keep hot things warm. Just be sure not to plug in and switch on your cool box…

OK, this is not the most crucial car hack, but you've got to admit it's a handy one!

CHAPTER 2

EXTERIOR HACKS

Your car may be your pride and joy but it doesn't take long before it's more of an eyesore. Car park dings, tiny scratches, cracked and faded plastics and foggy headlights may be invisible to a stranger's glance, but once you've seen them they cannot be unseen, and they all add up to make you feel gloomy about something that should fill you with joy. If anything, it's even worse under the skin: unwanted mechanical noises, smoky exhaust and sloppy, saggy controls combine to make your motor feel tired and old before its time. This chapter will give you some simple jobs to help you fall in love with your car once more, plus some useful tricks to get you out of a tight spot.

Maximise your car's life

Whether you own a classic car or have a personal contract purchase (PCP) agreement on a new one – or anything in between – you'll want to keep it running its best and ensure that it lasts the distance. The PCP car needs to go back to the finance company in perfect order, or else you'll be hit in the pocket. And you may want to own it at the end of the lease, so it just makes sense to treat it right. The classic car ... well, it's a classic car. You cherish it for reasons other than the practical.

This section will list the best ways to make your car last longer: some may be obvious, some will surprise you. Some are 'hacks', and some are just good pieces of advice that shouldn't be left unsaid. But it should be noted that these are things you should do over and above regular servicing, maintenance and checks – get hold of the Haynes Manual for your car to discover how to perform those tasks.

DRIVING

■ Don't warm up the engine at idle before setting off. There's no cool air coming in through the vents so the engine isn't operating efficiently. This is not good news for the internals as the inefficient combustion cycle can lead to various deposits, resulting in shorter component life.

■ Stick to low revs when you start up, particularly in the cold. The engine is at its most vulnerable when cold so this will ensure maximum longevity.

■ When setting off, take it slowly. The running gear suffers the most wear in the initial ten minutes of a drive.

■ Don't sit at the lights with the clutch pedal down or hold the car at the biting point as you wait for the lights to go green.

■ And similarly, use the brakes to slow the car down, not the gears. Both will wear down the respective components, but brake pads are much cheaper to replace than the clutch!

■ It seems obvious, but don't accelerate quickly and take it to the red line every time you set off if you want to preserve your car's mechanicals. And certainly avoid doing this in extremes of temperature.

■ Avoid holding the steering wheel on full lock for too long, as this will wear the hydraulic power steering pump.

■ Drive during low traffic times to reduce stop-start situations and long periods motionless at idle speeds. Travel planning and navigation applications like Google Maps and Waze can inform you in advance when the high traffic times are on your projected route.

■ A thirsty car is an unhappy car, so calculate the fuel economy of your car and don't rely upon the in-car trip computer (if it has one). Either use an app, Road Trip, Fuel Monitor or any of the other various fuel and mileage trackers, or simply keep a notepad and pen in the glove compartment and use them to record your fuel and mileage.

STORING

■ Before parking your car for a period, fill up the fuel tank to help prevent condensation build-up.

■ Thoroughly wash, polish and wax the car to protect the paintwork throughout its storage period.

■ Place a small chemical dehumidifier in the car – the rear parcel shelf is a good position.

■ Raise the car on axle stands to take the weight off the suspension.

■ Disengage the parking brake to avoid the brake sticking.

■ Place the battery on a trickle charger, ideally disconnecting and removing the battery first, particularly if planning to store the car for a long period.

■ Plug the tailpipe with a rag to prevent damp from getting in, but remember to remove it when starting up – leave a reminder note on the steering wheel.

Maximise your car's life

GENERAL

■ If you don't have a garage, or are out and about, aim to park in a shady spot. This will minimise interior damage from sunlight and heat. Try to avoid trees because they will often deposit sap or bird guano. If you have trouble finding good shade, get hold of some reflective car shades to place on the windows to minimise the effect of the sun on the car interior.

■ Don't carry too much unnecessary load and never exceed your car's roof load specifications or weight limits. Overloading leads to inefficiency and premature wear and tear, at the very least. In worst-case scenarios it could damage vital components or dangerously increase the vehicle's instability.

■ Keep the door and window seals from age-related cracking and damage by wiping silicone-based lubricant or a specialised rubber protectant on them periodically. These shouldn't need the frequency of protection that the paintwork or interior requires, so every month should do. Don't use petroleum jelly like Vaseline on these because it will damage this delicate rubber.

■ Clean under the bonnet – a clean engine will run cooler and therefore more efficiently than a dirty one. Also, if it isn't grubby in the engine bay it will be easier to spot fluid leaks. Protect the air intake, distributor and electrical parts before you start – you can use old plastic bags. Gunk is the renowned brand in engine cleaners, but if you're doing it regularly you may be able

to use standard washing-up liquid and a hard-bristled brush to get rid of the worst of the grease and grime.

■ Regularly clean the car's upholstery, and use leather preservation cream to ensure the interior remains in good condition.

■ Place a protective plastic sheet and an old towel under baby seats to prevent damage to the upholstery from spills. This mustn't interfere with the security of the child seat, however.

■ Always have your car's wheel alignment checked after buying new tyres and when you replace suspension parts. This will ensure that your tyres are operating efficiently and will wear predictably.

Use nail polish to fix paint chips and scrapes

There are numerous things that can start you feeling negative about your car, from a messy interior to chips and scratches on the paintwork. It's the latter that we're focusing on here, particularly the very small nicks and scrapes that no one would spot at a glance but once you've spotted them on close inspection, you will see them every time you look at your car.

Touch-up paint that is an exact match costs a fair bit, but you might well have something that does the job sitting at home. A near-matching nail polish will smarten up small stone chips a treat, ensuring that your car passes the kerbside inspection test.

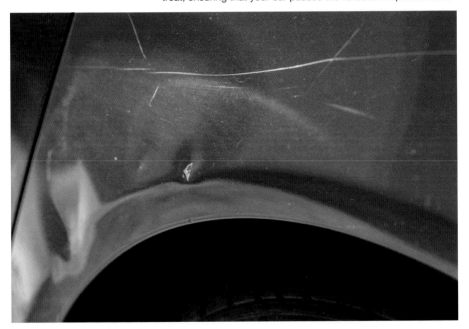

Use a plunger to fix dents

Serious dents will require either a full repair kit of slide hammer, filler and paint, or a trip to the bodyshop. There is no getting around that. However, smaller parking dings and some larger dents that haven't punctured the paintwork could be visibly improved or even fully mended with the help of this hack.

First of all, make sure your plunger is nice and clean, and free from jagged edges – you don't want to be adding any scratches to your car's paintwork. Then you want to ensure that the plunger gets good suction to the bodywork, and this is really why this hack will only work on a flat surface. If the dent's on a crease or corner, a plunger will not help.

To get good suction, dip the end of the plunger in a bucket of clean water or smear a very thin layer of washing-up liquid around the edge of the plunger's cup. Then gently apply the plunger around the dent – it's important that the cup is larger than the dent you're trying to repair – and give it a gentle pull.

It may take a few attempts to pop out the dent or it may work first time, but this hack should leave a smooth surface once again.

Use hand sanitiser to de-ice keyholes

This is a great hack for winter driving, as it utilises a household product that is often stocked up on when the temperature starts to get chillier. Hand sanitiser prevents those germs that appear just as the nights get longer from being passed around, and many of us not only keep big dispensers of it in the home but also carry pocket-sized bottles around with us to help give winter bugs a swerve.

The astute car hacker will be aware that hand sanitisers use alcohol to perform the anti-bacterial task they are designed for, and that alcohol can quickly reduce the temperature of ice owing to its lower freezing point than water.

Which brings me to frozen car locks, the bane of the winter driver (along with windscreen ice – more of which elsewhere in this chapter). If you discover you can't get the key into your car door on an icy morning, and the car isn't blessed with a keyless 'plipper', squirt some hand sanitiser into the keyhole and wait a few seconds for the alcohol to do its work. You should be able to slide your key in and get out of the cold in no time.

Rub out scuffs
with toothpaste

Toothpaste is an evergreen hacker's tool. You may have heard of its ability to remove scratches from CDs (remember them?), and this is mainly because of its properties as a mild abrasive. The fact that everyone has toothpaste in their home is what makes it so useful; it's always to hand.

This hack utilises that abrasive property to smooth out scuff marks from your car's paintwork. This works particularly well on marks that haven't scratched the paintwork down beyond the paint layers, although you can try it on various scuffs and scratches in the knowledge that the toothpaste shouldn't cause more damage than you already have.

Grab a tube of toothpaste and squeeze a small amount on to your finger, then rub it gently around the scuff mark. Then get hold of a drill and polishing wheel and carefully rub the area of the scuff until it has been blended into the surrounding paintwork. Wipe off the toothpaste residue, and you should have a smooth piece of paintwork as if nothing had happened.

When doing any work of this type on the car's paintwork, it's a good idea to follow up with a good wash, polish and wax to protect the paint.

Temporary wiper blade replacements

This hack is a temporary fix rather than a permanent one, but it's a tip that could help get you home in a situation where your windscreen wipers have become damaged or entirely worn out during a journey. The rubber blade on windscreen wipers is the perishable part, and while it will generally wear down gradually over time there are some occasions when the rubber will tear itself loose from the arm suddenly, usually while under heavy use.

Heavy use, such as during driving rain and spray, is of course when you need wipers most, so to suddenly have these visibility essentials rendered useless is highly dangerous. The dedicated motorist brought up on the *AA Book of the Car* will dutifully inform you that you should always carry a box of essential maintenance and repair items in your vehicle, and that this should contain a spare set of wipers, but in reality most of us do not have the space or (let's face it) the inclination to do this. There is a hack to get you home, however.

A woman's stocking or tights, when wrapped around the wiper arm, will stop the bare wiper arm or heavily damaged blade from scraping your window and should perform passably at clearing the windscreen of water until you can get to an auto parts supplier.

Temporary window damage fix using clear nail polish

This hack must be prefixed with a warning: large cracks in the windscreen are potentially dangerous and they should be repaired or the screen replaced by a specialist. Windscreen chips can also be a UK MOT failure: the maximum size for windscreen damage is 10mm in the driver's line of vision (equivalent to a vertical strip 290mm wide centred on the steering wheel), and 40mm elsewhere on the area of screen swept by the wiper blades.

Even small chips can lead to bigger problems. The chip weakens the screen's structure so a jolt from a speed bump or pothole can turn a chip into a crack or worse. Small stone chips are usually nothing more than unsightly, however, and that's where clear nail polish can assist.

First clean out the chip as best you can and gently dab it dry with a cotton bud, making sure not to leave any minute pieces of cotton in the chip. A compressed air line would do an even better job of drying out the recess. Now carefully paint some clear nail polish into the crack to leave it looking less obvious. This may even strengthen the area to help delay or prevent further damage.

Plastic bag door mirror ice preventer

Cold weather plays havoc with journey times, and that starts the moment you step outside to prepare your car for setting off. Snow, frost and ice equate to time spent clearing the car's windows, adding aggravation and leading inexorably to the frustrations associated with the feeling of running late.

Prevention is better than cure in every walk of life, and there are a few hacks that you can initiate as soon as the weather forecast indicates a serious drop in the mercury.

Everyone knows the tarpaulin over the windscreen trick to prevent snow and ice building up, and this remains a great way to save time defrosting windows in the morning, as well as helping your pocket and the environment by negating the need for de-icing spray.

But another less well-known hack relates to the door mirrors. These can be a pain to remove ice from, with plastic scrapers often being too wide to do an effective clearing job, and it can also be easy to forget to clear them of frost altogether – I've certainly driven off having cleared all my windows, then peered into a mirror at the first junction en route to find it rendered useless with ice build-up.

The solution is found in the kitchen drawer, (ironically) in the form of zip-lock freezer bags or equivalent. Find a couple of the appropriate size, larger than your car door mirrors. Slide them over the mirror housings and partially close the zip-lock so that they are as snug as you can get them.

Once removed they will have protected your mirrors from ice, and they can be stored in the car for reuse on the next occasion that Arctic conditions descend.

Use socks to protect wiper blades

Elsewhere in this chapter I remark upon the flimsiness of windscreen wiper blades. To be fair to the poor old wiper, they take a fair bit of a battering during their lifespan and none more so than when it's icy.

How often have you forgotten to flick the wiper stalk back to the off position when you park at night, then turned on the ignition on a frosty morning to be met by an agonising crunchy scraping noise as the wipers drag their way across the frozen windscreen? I'm sure it's not just me…

Wiper blades will freeze to the windscreen in sub-zero temperatures, and this hack will prevent that – and it won't matter if you've left the stalk in the wipe position. If you know it'll be icy overnight, before you head indoors for the evening just prop your wiper arms in the upright position and slip a long sock over each one. A zip-lock food or freezer bag will work just as well, but please remember to reuse them for this purpose rather than disposing of them (or using them for food!).

The next day, slide the sock off the wiper and once your screen is cleared of ice reintroduce the wiper blades to the windscreen. Tips like this will increase the longevity of a perishable part of your car, as well as helping to take some of the stress out of cold mornings.

East is East

On various occasions in this chapter, the hacks have related to the issues caused by freezing weather. Various tools from the car hacker's armoury have been utilised to good effect in the name of avoiding common problems faced in sub-zero temperatures, which, when combined with good motoring advice, should ensure that you don't have to stop when the temperature drops. For this hack, however, we're going to tap into nature's own suite of wonders.

The most annoying car-related task hands down is removing snow and ice from the windscreen. There are various ways and means to solve this problem, including scraping (time-consuming, taxing), using de-icer spray (costly, environmentally iffy), blasting the heater at the window (very time-consuming) and putting a tarpaulin or blanket over the windscreen the night before (awkward, requires prep, may leave you covered in snow).

This can all be avoided by parking your car facing east, if that's possible. As the sun rises in the east, it'll beam down on your car's windscreen before you've finished your morning coffee, and with any luck you could even come out to a completely clear screen by the time you're ready to set off. OK, that'll require a special set of circumstances, but at the very least it should make scraping quicker and easier. The sun may not reveal itself every cold morning, but if you remember this tip you'll at least be able to take advantage if it does.

TOOL HACK #3

Use two wrenches together for extra leverage

When trying to get a stuck bolt to budge, get some extra leverage by looping another wrench's ring or box end on to the spanner end of your first wrench. Bingo: instant extra torque.

CHAPTER 3

GARAGE HACKS

With a garage comes somewhere to work on and store a vehicle, preparing and protecting it from the rigours of outdoor life. Most people will stock their garage full of the necessary equipment to care for, diagnose, maintain and repair their car, and while there are ways for the car hacker to twist these tools for purposes beyond their intended use (see the tool hacks dotted around this book), this chapter instead continues the theme of the book by detailing how to use everyday household objects to complement traditional tools when completing tasks around the garage.

Protecting tyres from overspray

Painting repaired parts of the car can be undertaken in a well-ventilated garage, and this hack provides a helping hand when it comes to one of the more onerous tasks when undertaking localised spray painting – masking up to protect against overspray.

With some areas of the car, there are no shortcuts to masking up the areas where you don't want any paint to land. Using masking paper or newspaper attached to the car with masking tape is the only way to ensure you won't have to scrape overspray off the windows.

However, there are a couple of hacks to avoid paint overspray on tyres. Often, people who are undertaking a bit of home bodywork repair just let overspray land on tyres – they're perishables, after all, so they'll be replaced at some point. For the rest of us who want our car looking its best at all times, this is not an option. Tyres are a tricky area to mask up, but a pack of playing cards and a tub of grease or petroleum jelly will ensure your tyres are in the same condition that they were pre-painting.

Yes, playing cards. When painting a wheel, for example after repairing kerbing or salt damage, rather than masking up the tyre or removing it altogether, grab a standard deck of 52. Presuming that the wheel has been removed from the car and laid flat on the ground to undertake the repair work, tuck the playing cards between the rim and the tyre all around the circumference of the wheel. Then the paint can be sprayed on to the wheel without overspray bothering the tyre.

When painting bodywork near a wheel, for example after repairing notorious rust hotspots like the wheel arches, the tyres can be protected even when they're left on the car. Use petroleum jelly or, even better, thick grease to mask the tyres before beginning to paint – simply smear it generously on the tyre area adjacent to where the paint will be sprayed. The paint won't adhere to the greased-up surface, and it can be wiped off with a cloth afterwards. This hack is really a solution for very localised painting, because if a larger area is being sprayed it makes sense to fully protect the tyre and wheel with masking tape and paper.

Create a handy map of bolts and screws

When undertaking repair or modification projects that require dismantling a section of the car's interior, whether panels, door cards or even the full dashboard, the biggest fear is lost screws and other fixings. The difficulties that this can cause range from the minor inconvenience of having to replace them at the local parts store, to the hair-tearing and potentially fruitless search for rare and discontinued parts on eBay.

So it's best not to lose anything in the first place, but tossing all the bits into a plastic tub comes with its own issues – namely which screw goes where when it comes to reassembling. If there's one thing that's guaranteed when dismantling part of a car's interior, it's that nary two fixings will be the same. Why have six identical screws attaching a door card to the door when the manufacturer can use several slightly different ones instead? Reassembly becomes a lottery if the fixings are haphazardly gathered in a box.

The car hacker is not so carefree. Instead, you'll have kept hold of the sort of large piece of polystyrene that is used for packaging large electrical items such as televisions to assist with this screw-saving hack. You can also pick up polystyrene sheets from eBay or a DIY store for a few quid.

Using a marker pen, draw a rough outline of the item you're removing directly on to the polystyrene sheet. Then, as you remove fixings from the item, press them into the polystyrene board in the equivalent location that they came from on the item. For example, after removing a screw from the top left of a door card, puncture the screw through the polystyrene sheet in the top left area of the outline of the door card that has been drawn on the sheet. When reassembling, just refer to your polystyrene 'map' of the item you're working on to find where the fixing goes again – and the actual screws are right there on the map for you to use! Genius.

This hack can also be done with a large sheet of paper in a similar manner, and either pushing the fixings through the paper at appropriate places or sticking them on with a piece of masking tape. This is not quite as efficient as using polystyrene sheet, because the paper can tear and fixings may not adhere very well, but it can be done if you're desperate.

Magnet in breast pocket to keep tools handy

Here's a nice hack for when you're working on a vehicle, particularly in the engine bay or if the car is on a ramp. Quite often, the same tool is required numerous times while doing a single job, but both your hands are needed in between times. If you put a spanner or socket wrench down on top of the cylinder head or balance it elsewhere under the bonnet, it risks being knocked on to the floor, which can prove very frustrating.

Overalls, shirts and polo shirts with breast pockets should be ideal for this, but a long spanner will fall out of these shallow pockets as soon as you bend over the engine bay again. The solution is a flat, strong magnet. Pop one in your breast pocket and then you can attach your tool to the outside of the pocket, so it's always at hand while you're wrestling with an obstinate under-bonnet object.

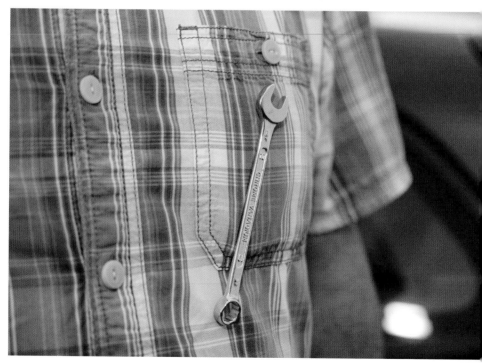

 ## Take photos of trim locations

This hack sounds like common sense, but that's quite often what car hacks are – common-sense ideas that perhaps you weren't aware of. Large areas of the interior trim of most cars are sectioned into smaller parts, or made up of more than one layer of pieces. Take a dashboard, for example – there are often various sections of the glovebox, centre console, heating surround, under-dash area, instrument surrounds and more. And when dismantling all this, to make a repair or for thoroughly cleaning and detailing, it can be easy to get confused when a large pile of interior parts are piled up outside the car.

The solution to this is to get your smartphone or camera and take a photo of each part in situ before you dismantle it. Then you can refer to the images when it comes to sorting through the parts that have been removed.

This has an additional benefit. Looking at the order of the photographs from newest to oldest will show you in which order to put the parts back together again. Even for relatively simple tasks this is a worthwhile exercise, because you could be interrupted midway through a small vehicle project for hours, even days and weeks, so the ability to refer to photos of the deconstruction phase could save a lot of guesswork.

 # Labelling wires when working on car electrics

Vehicle electrics is an area that intimidates most rookie car-tinkerers, but if you want to perform even some of the more rudimentary tasks on a car in the garage, chances are you'll have to do some form of electrical work. On most tasks this is likely to take the form of disconnecting some connectors and reconnecting them again when you've finished the job. Sounds simple, but when there's a spaghetti of wires and cables hanging out of the back of something you need to detach, people can go running to the nearest auto repair shop instead of rolling up their sleeves and getting stuck in.

Haynes Manuals contain wiring diagrams, which are an indispensable guide while doing any electrical work, but you can make things even easier by using the diagrams in conjunction with this little hack. When you unplug a connector, fold a small piece of masking tape around the wire and write what the wire is on the self-made label. Use a number, code or write exactly what it does – whatever works for you. Then you'll easily be able to reconnect everything correctly during the reassembly stage.

Find water ingress

Owners of older vehicles will probably be aware of that crushing feeling of finding water inside the car. Mopping it up and drying it out isn't a major problem, and fixing the leak is rarely a huge undertaking. However, sourcing the origin of the water ingress is a major headache. This hack is more of a series of checks to use as reference should a puddle form in your car's footwell.

Before starting, it's important to find out whether the water appears after rain or not, to discover if the water is from an external source or from part of the car itself.

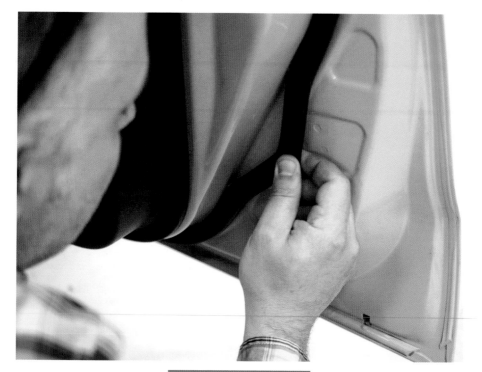

DOOR SEAL

The first place to check is the seal around the door that is nearest to the damp area. These can get worn and cracked, meaning the door won't seal tight and water can creep into the car. Repair is a simple case of replacing the seal – this is not a costly part. If the seal looks in good condition, perform a test with a hose (not a pressure washer!). If water is still getting in, the door may not be fitting into the aperture properly. Has the car been in an accident, even minor? The door may have bent outwards slightly and come away from the seal. You can try to bend it back (with the window rolled down), but this may require bodywork repair or a second-hand replacement door. Application of duct tape may do the job of meeting the seal if the gap isn't large – it's worth experimenting here.

DOOR CARD

DOOR WEATHERSTRIP

A common source of water ingress is the door card. Rain can enter here usually via the window, but it's designed to run down a membrane between the interior door card and exterior door skin and then exit through drain holes at the bottom of the door.

However, the membrane can get worn or torn, and that means water can soak into the door card. This water will collect in the footwell, so if you find a puddle there it's worth checking the bottom of the door card for dampness. The repair here is simple – remove the door furniture, dry it all out and repair the membrane with waterproof tape.

The flap that sits at the bottom of side windows is designed to keep most water out, but it should be tight to the glass otherwise it'll allow rain to pour in, with the inevitable consequence that some will enter the cabin via the door card. Take a close look at it and rub your finger along – if the gap's too large, consider replacing it.

A less common area for water ingress is the windscreen rubber, but this is worth checking if all else has failed, especially if you have recently had a replacement windscreen – as it may not have been fitted properly.

SUNROOF

The sunroof is another part of the car that is designed to let in a little water, but it's then channelled through gutters and guided out of the car via drain holes. However, these drains and gutters can get bunged up over time, and this means the water will attempt to head elsewhere – often into the cabin. The only real way to test this is to pour water in and closely inspect whether it drains away or pools up in the drainage channels. That'll give you a clue to the culprit. This repair involves a bit of detective work – try to find out where the drainage channel goes and blow the blockage out with an air line.

POLLEN FILTER

Pollen filters are usually located under the facia in the passenger cabin, but are sometimes found, especially in older cars, concealed in a large rectangular plastic box positioned just ahead of the windscreen under the bonnet – you may have to remove some trim at the back of the engine bay to access it. If this is the case, these can leak if the lid is ill-fitting or damaged, or if the box has cracked. It's probably worth replacing the box if it's damaged rather than taping up any damaged parts.

HEATER MATRIX

A common problem on older cars, this is not actually water ingress but an antifreeze leak. The effect is the same, however – damp footwells. The heater matrix is a small electric radiator that water passes through to help heat the cabin. The matrix will rust over time, and this can lead to leakages, as can perished rubber hoses and loose or damaged connections. Repairing the heater matrix is a big job involving removing the dashboard, so if you're not confident then get an expert in.

AIR CONDITIONING

Air conditioning units are designed to condense and drain off water – you will sometimes see a small pool under the car after using it. However, just as with the sunroof, the drain tube can become bunged up and the water reroutes into the cabin. The best way to diagnose this is to mop up and dry off the damp area, then don't use the air conditioning for a while. If the water ingress stops then you have your perp.

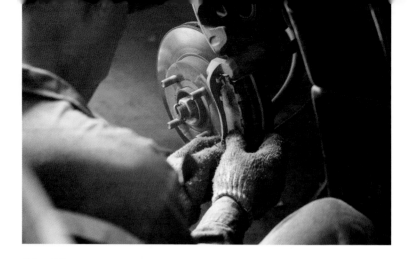

Bungee to hold brake caliper

This book is not aimed at amateur mechanics above a very basic level of knowledge and inclination to work on their car, but having said that working on brakes is something that anyone should be able to do with a bit of assistance from a Haynes Manual. This hack is more of a mechanic's tip; a piece of knowledge that someone starting out on car maintenance may not be aware of. And it also uses a great car hacker's tool: bungee cord.

I'm not going to go through the process of changing a brake pad, because details of that will be model-specific. The basic process is the same for all disc-brake cars, though, so here is a brief run-through to help explain where the hack comes in.

Loosen the bolts of the wheel you need to remove, lift the car with a jack and place axle stands under the frame rails. Fully remove the wheel bolts and wheels. Carefully remove the retaining clip on the left front caliper by levering it off with the screwdriver. Remove the caliper bolts. Carefully pry the caliper off the pads and rotor with the screwdriver, but at this point the caliper will only be attached to the car via the brake hose – which is not advisable.

This is where the bungee cord comes in. Thread the bungee cord through the caliper and hang each hooked end of the bungee to the frame or coil spring. Now you can continue with the brake pad replacement: slowly compress the piston of the caliper inward using the C-clamp until it bottoms out. Remove the old pads and use a stiff wire brush to clean the surface of the caliper anchor where the pads were. Apply a coat of silicone brake lubricant to the cleaned area, then place the shims on the pads, install the new pads into the caliper anchor; and reassembly is the reverse of removal, to coin a phrase.

Prevent door damage with pool noodles

One of the issues of single garage ownership is that most garages are quite narrow, and this issue is particularly prevalent with the average size of the family car increasing in length, height and width every year. This lack of sideways space can play merry havoc with the outer edge of a car's doors, as it's highly likely that on more than one occasion they'll come in contact with the garage wall, scraping the paint from the door edges and leading inexorably to corrosion.

Door edge protectors could be purchased, but let's face it, even the most subtle of these plastic door-edge strips is an ugly accoutrement to a car's profile. Fortunately there's another solution that'll leave the door edges as the car-maker intended, and it uses pool noodles.

Pool noodles are also known as pool or swimming woggles, and they're those long cylinders made of thick foam that people, mainly children, can wrap under their arms and around themselves to aid flotation when they're learning to swim. For the car hacker, they're much more than that.

Attach a pool noodle, cut in half along its length, horizontally against the wall on each side of your garage, using the appropriate bolt and wall anchor for the wall material, positioned in line with the doors and at the height of the widest part of the door when it's open. Now when your door is opened the woggle will provide a cushioned bumper, meaning the door edges are protected from scrapes. And if your garage is shallow, a noodle or two could be fixed vertically to the far wall to protect the bumper too.

Hang a tennis ball at the back of your garage roof as a parking guide

This is a hack for cars from the pre-parking-sensor generation, of which there are still plenty around. It can be problematic gauging the distance between the front and rear extremities of a car when parking in a garage – they're often dark and the dimensions are tight. There's a great hack involving a tennis ball and a piece of string that can help out, though.

Park the car in the correct position in the garage, then measure the distance between the garage ceiling and the centre of the car windscreen (or rear screen, depending on which direction you prefer to park in the garage) using a length of string. Mark that position on the ceiling and add around 30cm to the length you measured, then cut the string to the resulting length. Now you can move the car out and attach a hook to the ceiling on your mark, then tie one end of the string to it. Thread the other end of the string through the tennis ball and tie a knot in the end to ensure it doesn't come off.

Test it out, and adjust the height of the tennis ball by moving the retaining knot until it's positioned against the centre of the screen. Now when you're parking, as soon as the tennis ball nudges against your front or rear screen the car is positioned correctly.

Use your phone's selfie camera to visualise awkward access points

The inspection mirror is arguably one of the most underrated of mechanics' tools, allowing you access to areas of the engine bay, underside and more that would otherwise be inaccessible without getting the car up on a ramp – and sometimes even then. But the car hacker has something that is arguably even better to hand: the smartphone. The smartphone selfie camera, and its related telescopic appendage the selfie stick, are, depending on your point of view, somewhere between a bit of fun to responsible for the collapse of society as we know it. The car hacker cares not for these judgements and only sees their usefulness in the garage environment.

Scrutinising the underside of the car's mechanicals is a doddle with the selfie camera: simply hold the smartphone as if it were a mirror and you can peek at difficult-to-access areas under the bonnet or beyond. Even more usefully, set the smartphone to record video of the area and you can review it later, zooming in if necessary and even getting a better view on a larger screen if required. This procedure lends itself to the use of a selfie stick, for example if you're trying to diagnose a problem under the car – set the camera to record and poke the stick underneath to get a great view of the issue with no axle stands or car creeper required.

One final tip, though: slide a powerful LED torch or lamp under the car too, because most selfie camera video settings don't have a flash.

TOOL HACK #4

Use WD-40 to clean oil from your garage floor

We all know WD-40 has hundreds of uses, but did you know that it can clean up any oil stains on your garage or driveway? That's on top of removing rust, loosening bolts, removing stickers, lubricating, cleaning bugs off your lights and windows, and much much more…

CHAPTER 4

TRAVEL HACKS

Travelling in the car, whether alone, with friends or as a family, on one-off journeys or a driving vacation, can be great fun. Making the journey part of the adventure is a great way to stop travelling by car from feeling mundane and a necessary evil. But it can also be stressful, frustrating and infuriating … which can also mean dangerous and costly. This chapter concentrates on a variety of travel-focused hacks that are designed to make your car work with you and not against you, increasing its usefulness and efficiency with the end goal of making a long journey a calm, enjoyable and more frugal experience.

Maximise your car's efficiency (hypermiling)

Long journeys are costly; cars are costly. Fuel, wear and tear to perishables and high-priced food outlets en route are just some of the things that suck your finances dry while travelling. These are the unwritten covenants that we buy into when motoring, and for some they're accepted without question. Not the car hacker, however – there are ways and means to squeak out every last penny from your car, and on long journeys that can add up to significant savings; money that can be spent on more enjoyable stuff when you reach your destination.

Having said this, it's important not to become too obsessed by making journeys cost-efficient. This can lead to stress as the feeling that other drivers are obstructing you from achieving maximum efficiency takes hold, and finding that balance between money-saving and in-car calmness is key to what car hacking is all about – not being cheap, but being cheerful.

The concept of 'hypermiling' has been around for more than a decade, and while early versions of this efficiency-at-all-costs theory included ideas that are just plain dangerous – such as maintaining a steady cruising speed even when going round tight corners – a refined version of hypermiling is a great basis for solid efficiency hacks. Let's look at these now.

KEEP TRACK OF YOUR FUEL CONSUMPTION

As outlined in the section relating to increasing your car's lifespan, logging fuel consumption helps you to get a feel for your car's economy – allowing you to adjust your driving style to improve it. Either use a mileage tracking app or a pad and pen to record your fuel and mileage, and don't rely on the car's trip computer.

ADD NOTES ABOUT YOUR OWN EMOTIONAL WELL-BEING

As well as recording details about fuel fill-ups, jot down notes about how you feel you were driving in between visits to the petrol station. If you notice that a period of poor fuel consumption correlated with times that you were stressed or driving angry because of traffic or 'bad' drivers, then you can try to make life adjustments to lessen the chances of this happening again.

ACCELERATE GRADUALLY

The less you open the throttle, the less fuel you use. Practise gently squeezing the accelerator until the required speed is achieved, rather than stamping on the pedal. If your car has variable speed cruise control, it can be used to slowly increase speed in small increments.

INCREASE YOUR STOPPING DISTANCE

Efficiency while driving is about ensuring that none of the car's momentum is wasted during a journey, and fundamental to this is leaving the right amount of space between your car and the car in front. If you're too close, you'll be forced to slow down and speed up as they do, and even use the brakes frequently and unnecessarily. Leave a very large gap and there'll be ample room to coast to a stop and gently accelerate through all manner of manoeuvres.

It may be that the car behind isn't interested in adhering to your hypermiling philosophy, but if it overtakes you try not to rise to the bait, and instead simply coast to a good-sized stopping distance behind the overtaker.

BE AWARE OF DISTANT HAZARDS

Traffic lights, junctions and motorway congestion are the types of hazards that are unavoidable but arrest momentum. The way to ensure maximum efficiency is to be alert to what's coming ahead. If you see a red light in the distance, take your foot off the accelerator immediately so you can coast up to it. Even if you can't see the hazard ahead, paying attention to when cars are braking up ahead can give you a clue to what's around the next corner. And in long tailbacks, try to find a slow, constant speed to minimise the accelerating, decelerating and time stopped that the ebb and flow of congestion causes.

CHECK TYRE INFLATION BEFORE EVERY JOURNEY

Tyres that are inflated to the maximum possible pressure for your car are operating as efficiently as they can, and part of that efficiency is reducing the contact patch on the road, which reduces resistance on the car.

AVOID PUDDLES AND POTHOLES

Puddles, bumps, cat's eyes and potholes will decrease momentum, increase drag and upset the car's efficiency, so they are to be avoided. If you are keeping a good distance from the car in front and being aware of upcoming hazards, they'll be easier to spot too. If you are on a regular commute, make a mental note of where potholes are and where puddles regularly collect, and avoiding them will become natural.

Maximise your car's efficiency (hypermiling)

WINDOWS AND AIR CONDITIONING

Open windows and sunroof cause turbulence that affects a car's aerodynamic efficiency, in turn reducing fuel economy. The air conditioning unit draws power from the engine and reduces efficiency that way. In an ideal world, you would close all windows and switch off the aircon, relying only on the air vents and a bottle of cold water. If the temperature really is unbearable, research suggests that open windows are more efficient than air conditioning up to speeds of around 55mph, when the opposite becomes the case.

MAINTAIN YOUR CAR

Keep up with regular maintenance – the Haynes Manual for your car make and model will help here – and following the tips and hacks outlined in Chapter 2 designed to make your car last longer will ensure that your car is running at maximum efficiency.

GAMIFICATION

A great way to ensure that you're getting the best miles per gallon out of your car is to gamify your journeys. Creating a game out of anything from language-learning to exercise and dieting has become increasingly common, aided by various websites and apps, and there's no reason why you can't do the same with hypermiling.

Using the data from your car's trip computer, or the recordings that you're logging manually, you can compete against yourself to get the best possible mpg on a journey you repeat often – the commute is perfect. It can be strangely satisfying, even celebratory, to beat your fuel economy record by 0.1 of a mile per gallon! You can even compete with workmates to find out who can improve their economy most over a set period of time.

Cupholder expander

Sometimes the cupholders provided by car manufacturers just aren't up to the job. They have often been designed to fit in a specific space in the car, not to allow decent-sized cups to sit inside them. This can be incredibly frustrating – I remember when my little boy had a particular water bottle that he used (and he wouldn't use any other!), and it was too wide at the base to fit into the cupholder next to his car seat. So every time he wanted a drink, whoever was in the passenger seat had to turn round and pass him the bottle. And if there was no one in the passenger seat … well, things got a bit frantic.

Fortunately I discovered a beautiful hack that can expand cupholders to fit all manner of outsized receptacles. Measure up your cupholder and head along to the plumbing department of the local DIY store. There you'll find PVC reducing couplings, or reducers. Select one that has a narrow diameter end that fits in the cupholder, and you can now fit larger cups and bottles in the larger end. If it's too short to be stable, buy a small piece of PVC pipe to attach to the narrow end, cut to size so it reaches the bottom of the cupholder.

Photograph car parking location

This travel-related hack is relevant to a midpoint in a journey, specifically when parking in an airport car park. It's designed to ensure that you don't go through the ordeal of forgetting where your car is parked when you return. It's easily done: tiredness, jetlag, forgetfulness and distraction all play their part in addling the brain when it comes to pinpointing the car's position in that bewildering ocean of vehicles.

This may seem like quite an obvious hack, but on arrival in the airport car park take out your smartphone and photograph the spot where you've parked. Some airport car parks have signs at the bay telling you exactly where the car is; others have signs marking the level, area and/or row. Take a photo of those, and at the very least you'll know which part of the car park to head to.

Tablet holder hacks

Keeping children entertained is the holy grail of long family car journeys, right? If they are happy, or distracted, for the full distance then it's like one of those adverts where the impossibly white-teethed parents grin their way through 30 seconds of beautiful weather and clear roads, weaving gracefully towards destination utopia. Let boredom set in and once the dreaded words 'are we there yet?' drift from the rear seats to the front, the next infinity hours of driving are more likely to be akin to a still from an apocalyptic zombie movie.

It used to be the height of luxury to have TV screens mounted in the back of the front headrests so that the chauffeured rear-seat passengers could inject a bit of first-class air travel into their commute, but nowadays many households have tablet computers or large smartphones that can temporarily do that job if attached well. There are tablet mounts available to purchase, of course, but such single-purpose expenses are not for the car hacker.

A couple of those evergreen hacker's friends, bungee cords, do just the job. Wrap one around the headrest legs and use the hook on each end to rest the tablet in. Then get the second bungee and wrap it round the first bungee cord, using the hooks to attach to the top of the tablet. This should stabilise it enough to stop it dropping out of the makeshift bungee clamp that's been created, unless the driver's performing some particularly spirited manoeuvres…

A large smartphone can be used in much the same manner, and instead of bungee cord all you'll require is a sturdy elastic band. Wrap the band around both headrest legs again, and then around either side of the phone. It'll hardly be a cinematic experience, but married to a pair of headphones it'll be enough to keep a rear-seat passenger occupied for a good chunk of a journey.

There is one final tablet holder hack which is a fail-safe makeshift solution when all else has failed: use a large, clear freezer storage bag. Drop the tablet into the bag, remove your car's front-seat headrest, drape the top end of the bag over the holes on the top of the seat, and replace the headrest using the legs to puncture the bag and secure it into position. You should be able to use the tablet's touchscreen through the bag, and you can puncture another hole in the side of the bag to thread the charging cable through.

Use a Magna Doodle as a combo car toy and food tray

Another tip for parents of young children here. Magnetic sketch toys, most famously known as Magna Doodles, are incredible journey toys because they're great fun, mess-free and will easily hold the attention of all but the most demanding toddler for a decent distance between comfort stops. They also don't use any batteries or need to be plugged into the 12v socket, so cannot run out of juice mid-doodle.

Even better news for the car hacker is that they have more than one purpose. On a food stop, there's no need for a temporary table – just use the toy as a tray. It'll do a decent job of keeping crumbs and spills off clothes and seat fabric, and can be wiped after mealtime to continue the fun.

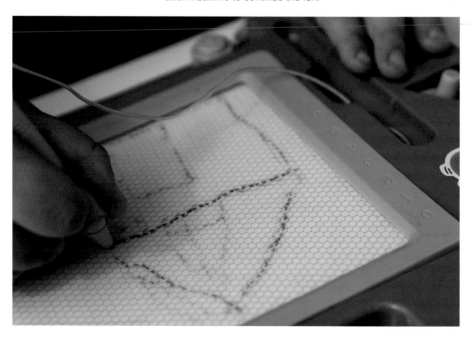

Wipe-off markers for drawing on windows

Children need constant attention on any journey, so I hope kiddie-free car hackers will excuse further suggestions for the child's plaything oeuvre. This tip carries a mild risk, however, as it involves giving the little ones access to pens in the back of the car. The pens involved are either dry-erase markers or paint/chalk pens – the type of pen that some shops use to paint special offers on their windows or on A-boards.

Bringing a dry-erase whiteboard seems sensible here, but they're quite large even in compact form, and if the car's already full up with holiday kit you probably don't want to add a large tray to the where-do-I-store-this? list, particularly because it'll probably only keep the little 'un entertained for a short time.

Wouldn't it be more fun to draw on the window? Adding their own touches to an ever-changing backdrop is tremendously entertaining for the huge imaginations in those small heads. Bring along a chalk duster or window scraper in the glovebox and the canvas can be wiped clean at every comfort stop. And if they get on the seat fabric, they're not too taxing to clean.

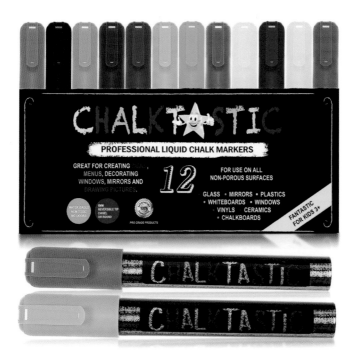

Snack packs

Everyone gets grumpy on an empty stomach, and during a lengthy journey the additional frustrations and mild claustrophobia of being contained in a car only intensify the feeling. Carrying a lunchbox of snacks is an obvious way of helping to deal with the potentially long gaps without a food stop, but what if you run out? A great hack is to carry large quantities of tasty dry food that won't perish en route.

What is required for this is a few large, resealable, wide-mouthed plastic bottles or tubs, for example the type that comes with infant formula, family- or catering-size packs of coffee or hot chocolate, and more. Fill two or three of these up with Cheerios or similar snackable breakfast cereal, small cheesy crackers, mini-pretzels, or any of the family's favourite dry snacks. They won't take up too much space, and will last more than one journey without going off.

This hack also encourages the reuse of habitually single-use plastics, so even the environment is a winner.

Make a glovebox kit

When you're a long way from home it pays to have some security in case something goes wrong. In a break from most of the hacks in this chapter, this outlines a useful thing to carry around in the car, especially if you're travelling long distances alone. Putting a pack in your glovebox that contains essential information and kit that you may require if lost, broken down or stranded could help you get back on track when you most need it. It'll also keep all that documentation in one place rather than spread between the glovebox, home filing cabinet or (more likely) odds and ends drawer, so you always know where to look when you need access to your car's paperwork.

An A5-sized sectioned wallet folder should fit nicely in the glovebox, and you can fill it with essential and useful things. The contents could include:

- Insurance details
- Breakdown cover information
- Your car's V5C document
- Checklist of essential information, e.g. tyre pressures, such as you'll find in every Haynes Manual
- Some emergency cash and/or a prepaid credit card
- Emergency contact information
- Slim portable charger (also known as a power bank) for recharging a smartphone

You may think of other useful things to carry in your glovebox pack, but it's worth noting that if you don't have a lockable glovebox then it's good advice to bring the pack indoors for safe keeping when you leave the car anywhere.

Avoid road rage

Car hacking is all about improving your motoring life, and using good advice and household objects to give yourself the best chance of a relaxed and happy journey. Tied up in this is the spectre of road rage: the way you drive, how others around you are driving and more importantly how you perceive them can have a devastating effect on your emotions behind the wheel. These tips should help you to avoid creating a tense environment inside your vehicle and will hopefully avoid any conflict while travelling.

EVERYONE IS HUMAN

The first and most important thing to remember when getting into a car is that all the other drivers are just like you, going through life's ups and downs, and are fallible. Everyone makes mistakes, and they're not winding you up deliberately when they drive slowly or quickly, or miss a junction, or have to stop to deal with their kids. Drive with empathy and you're halfway to avoiding road rage.

DON'T DRIVE EMOTIONALLY

When people are angry behind the wheel, it's likely that they're having to deal with some other emotional trauma in their life. Try not to drive immediately after a highly charged emotional situation, such as an argument or a break-up. Go for a walk and give yourself time to calm down before getting in the car.

BE PREDICTABLE

When driving, be aware of and alert to other drivers. The old Highway Code maxim of mirror–signal–manoeuvre is key to predictable driving, and if you stick to that then you're unlikely to provoke rage in those driving around you.

WE JAMMIN'

If you find yourself in unexpected congestion when you're trying to get somewhere, stick on some music, a favourite radio station, or a podcast or audiobook, and try to come to terms with the fact that you're going to be late. Phone someone and tell them if necessary – as soon as you hear someone's voice you'll realise that this is OK; it happens to everyone.

ALWAYS LET SOMEONE IN

If someone's waiting to be let out of a junction, let them out. Doing this will make the driver immediately in front of you a happier driver, and it makes you feel good too. And when merging lanes, get involved in the 'zipper' (everyone lets one person in) so no one's left waiting too long – it'll ensure good traffic flow, and that will make everyone happier.

MUSIC SOOTHES THE SAVAGE BEAST

Before setting out on a journey, create a playlist containing music that's guaranteed to calm you down or fill you with good feelings. Make sure your favourite albums are in the car or downloaded to your smartphone. If you feel the blood rising during a journey, stick on some soothing sounds.

AVOID ERRATIC MOTORISTS

If someone's driving erratically, don't stick close to them through curiosity or for your entertainment – just avoid them.

DON'T MAKE THINGS WORSE

When someone wants to merge into your lane, don't close the gap and give the 'stare-ahead', just let them merge in. If an impatient driver swerves out to overtake you, don't speed up to make it difficult for them. These things only make matters worse, threaten to create a road rage incident and can put you in danger of an accident.

BE THE BIGGER PERSON

Sometimes people will just be angry, and will honk their horn at you, shout obscenities or make aggressive gestures. Don't react. Stay calm. Don't do the same thing back but also don't smile or laugh, as this may wind them up further. Be the bigger person, and give an apologetic wave even if you think you've made no mistake whatsoever. You can relax in the knowledge that you've defused a situation.

TOOL HACK #5

Use a rubber band to grip a stripped screw

This doesn't always work, but placing a strong rubber band over a screw head can help to get those pesky stripped screws out. If that has no effect, weld another screw to the stripped head and get it out that way.

USE THE HORN AS AN ALERT ONLY

Speaking of the horn, only use it if you must, and only to alert other drivers of potential danger or if they've not seen you. It isn't helpful to honk the horn to show displeasure at another car's manoeuvre, and it could turn a bit of erratic driving into a road rage incident.

DON'T GO HOME IF FOLLOWED

If you witness a road rage incident, don't approach it – instead report the incident to the authorities. And if an angry driver starts following you because of some perceived feeling of injustice, don't drive home; you don't want them to know where you live. Call the police from a hands-free phone and drive to the nearest police station – if you don't know where that is, satnav will help.

CHAPTER 5

STORAGE HACKS

Cars come with a lot of space inside but they're not necessarily adorned with a multitude of clever storage solutions. While the advent of the MPV and the minivan led to some clever ideas, like under-seat drawers, recessed roof storage, myriad cupholders, coin slots and bottle bays, many areas have remained exactly the same over the decades since their introduction, like the glovebox, door pocket and boot space. The car hacker thinks more can be done with these areas to customise them to meet the owner's particular demands, and this is what will be covered in this chapter.

Add a curry hook

The curry hook has usurped the now ubiquitous cupholder as the most awe-inspiring, why-didn't-they-think-of-that-before, best-thing-since-sliced-bread element of a new car's interior. 'Oh look,' new-car buyers cry as they inspect an interior in the showroom, 'it's got a curry hook! Where do I sign?' Well, perhaps that's an exaggeration, but there's no doubting the usefulness of a small plastic hook sprouting out of the centre console in the passenger footwell. Whether it's for the transport of the bag of takeaway Indian food – the source of the device's name – and stopping a korma from ruining the upholstery when sweeping round bends on the way home, or to keep any other bag handy and upright, the curry hook has a practical benefit.

It will come as no surprise that a simple hack is possible in order for one of these divine objects to bring its perks to your car. Strong self-adhesive hooks are available in various sizes from most DIY stores, and while most are white you can find darker ones too that will make a better match with car interior plastic. Simply find a suitable spot on the centre console in the passenger footwell, clean the area with spirit and stick the hook on according to the manufacturer's instructions.

The perfectionist car hacker could colour-match the self-adhesive hook to their car's interior with a combination of plastic primer and spray paint, or even get it plasti-dipped. And it's worth mentioning that you can add that top hacker's tool, the good old binder clip, to the curry hook and carry things that don't have handles too.

Make boot dividers and add boot storage

The car boot is the most obvious bit of storage space in a car, but its cavernous nature is both a blessing and a curse. When transporting large quantities of gear it's fantastic, and the more van-like the space is the more stuff can be crammed in. However, when carrying few items, or a single bag of shopping, it's not too helpful – driving the car, no matter how carefully, will toss the items around the boot's expanse.

Boot dividers are the answer, and many manufacturers offer tailored boot-sectioning devices as option at point of sale or as aftermarket extras. However, these are often extortionately priced, no doubt because they're tailored to one specific model. Tailoring isn't necessarily required, though. Here are three household objects that can be used to hack some perfectly workable boot dividers.

LAUNDRY BASKETS

Laundry baskets are ideal for collecting various small objects in or a couple of shopping bags. Putting in as many as will fit in your boot will stop any chance of them tipping over, but to be honest you'd have to be participating in a round of the World Rally Championship to tip all the contents out of one.

SHOWER CADDY

A portable shower caddy – those sectioned plastic tubs with a handle in the middle – are ideal for ferrying individual bottles and smaller objects. For example, car spares could be deposited in here, such as engine oil and other essential fluids, spare bulbs, tyre inflator and lithium battery charger.

IKEA DRONA

Anyone who isn't aware of the storage revolution provided by Swedish furniture-maker IKEA's Expedit (and later replacement Kallax) shelving units has probably been living in a cave for the last 30 years. The cuboid shelf spaces are designed to be filled by various IKEA drawers and boxes, and the Drona fabric storage boxes are exactly the sort of large and deep receptacles that fit our purposes here. They're probably a better solution for hatchbacks: simply fit as many as you can into the boot space. They act as great sectional space, and the built-in fabric grab handles make them a doddle to lift in and out of the back of the car.

Install a semi-permanent cool box in the boot

This is a neat trick to add a bit of camper van-style flexibility to your car. It should be noted that this is really a hack for hatchbacks, estates or SUVs, as a booted car won't allow the access that's required to make it work. Cool boxes are incredibly useful not just for holidaying but also for a trip to the shops or on a day out. But they take up rear footwell space if they're used in the cabin, and they can get tossed around the boot when it's otherwise empty.

One way to avoid the issue of boot-space instability is to combine one with the boot divider hack described elsewhere in this chapter. Essentially, wedge the cool box into an area of the boot with whatever you're using to divide up the boot.

But why not carve out a niche for the cool box? If it's to be a semi-permanent addition to the boot, here's a neat trick. Get hold of the boot liner for your specific make and model of car, and mark out an area next to an edge of the liner. Then screw a pair of wooden battens into the top of the boot liner, screwing from the bottom of the liner, to create a recess that the cool box can sit in without toppling over or sliding around the boot. For the best effect, paint the battens the same colour as the liner or your boot interior. But how to power the cool box? For this, follow the instructions elsewhere in this book on how to add 12v outlets to the interior of your car, adding one to a trim panel close to where you've stationed your cool box. Now you have the next best thing to a fridge in your car, adding some lifestyle utility without the need for a camper van.

Roof box hacks

A good roof box is exactly the sort of utilitarian accessory that gets most car hackers excited. Doubling the storage space of a car with a big, streamlined load carrier is an immense way to increase the flexibility of your vehicle's space. But roof boxes are not without their problems – they are in essence a big empty box with no clever way of storing items; they just supply the space and the rest is up to you. And that means that if the roof box isn't filled with belongings, they can move around, making noise and unsettling the occupants.

No problem for the car hacker. There are a few hacks to stop loose objects moving around a roof box, the first of which is to get hold of a mesh bungee net or a cargo net. Inside the roof box should be the mounting points where it attaches to the roof bars. The hooks of the bungee net should be able to attach to these, so you can then keep smaller bags, boxes and other items in the roof box restrained under the mesh.

Another way to stop smaller objects from irritating you when being carried by roof box is to fill the box with your chattels then fill the remaining space with an inflatable lilo or similar. That should wedge everything in place so it doesn't roll around, and you can simply deflate it once you get to your destination.

And finally, if you wish to carry many loose items in a box on top of the car, many of the boot-dividing hacks that you'll find elsewhere in this chapter will apply to roof boxes too.

Create a backseat organiser with a shoe rack

Cars have door pockets, boot space, seat-back pouches, armrest cubbies, gloveboxes and various other storage areas, so when you're travelling with children why does it seem like there's not enough room for everything? The hack here may not cure this logical impossibility, but it'll give you a fighting chance of storing everything you need when transporting infants, toddlers and even older children.

The hacking device required is a shoe organiser or two. These are the large hanging sheets that usually contain between 16 and 24 pockets for sliding shoes and sneakers into, which are traditionally hung from a door or inside a wardrobe or closet. They can be picked up from online auction sites, local classifieds and charity shops, and will usually cost a few pounds. You may even have some lying around.

Hang one from the back of each front seat, and strap the bottom to the base of the chair to avoid fouling the rear footwells, and instantly you have multiple storage pockets for everything from baby wipes and nappies to portable games consoles, books, travel games and snacks.

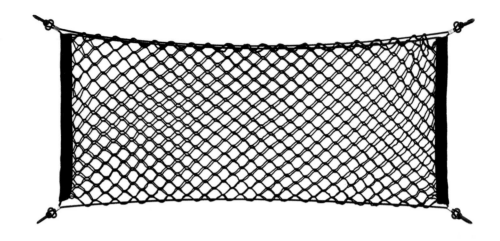

Mesh bungee net can add extra storage

If the bungee is one of the car hacker's favourite tools, the mesh bungee net must be the pinnacle of hackerdom. This strong, elasticated cargo net is used mainly to hold items to the floor, but for this hack it's going to be utilised to create additional storage above the occupants' heads. You can pick up a mesh bungee net for less than £10 from motorcycle supply stores, and they're useful for various applications so it's money well spent.

Attach each corner hook to one of the car's interior grab-handles, situated above the doors. Use a carabiner or even zip ties to secure more firmly, if necessary. Now the bungee net should be stretched across the interior roof lining, but stretchy enough to store objects in. Aim to store large but light items in there, such as winter coats, and these cumbersome pieces of luggage will be nicely tucked out of the way, leaving regular storage areas free to contain other things.

Utilising previously useless space

There are some areas of a car's interior that have to be left empty for practical reasons but end up being useless space. One such is the under-seat area and the space directly under your knees once you're seated. The latter must be left empty because the seat slides forward and back to allow for flexible legroom, but once you're seated there's a big empty space right under your legs.

Take advantage of that with some well-sized storage boxes. For example, IKEA sell reasonably priced fabric shoe boxes that slide partially under many car seat areas and poke out into that unused area under the knees. You can also pick them up on eBay and similar. Other suggestions I've seen for this storage area are covered cockpit organisers designed for small aeroplanes, and small tool boxes or craft boxes. The dimensions of the space to fill will vary from car to car and passenger to passenger, so it really is a case of measuring it up and going on the hunt for clever storage solutions to fill the gap.

Something out of muffin

There's been a lot of reference to cupholders in this book, and details of how to improve existing drinks receptacles, but what if there are no cupholders in the rear of your car? This hack provides a solution from the kitchen cupboard. It's a muffin tray – a baking sheet with recesses for muffin mixture.

Place the muffin tray between the two rear-seat passengers, and they'll be able to rest most takeaway beverage cups in the recesses. They'll remain stable while on the move – or certainly more stable than if they were placed directly on the seat.

Putting the muffin tray in a similarly sized laundry basket will add an extra level of security, and it'll now double up as another useful item for a journey with the whole family. When you stop to purchase lunch from a fast-food outlet, take the tray/basket combination along with you, and you can then securely carry back multiple drinks cups along with your food.

Tiny containers to divide up storage areas

Elsewhere in this chapter you'll find hacks for dividing up your car's boot space and roof box storage area. In this hack, I'm taking that theory and shrinking it down – a nano-hack, if you like.

Various storage areas in the car are empty spaces that would benefit from being compartmentalised. Centre console storage, such as the armrest storage bin, is usually a deep trough so, while it offers a reasonably large space, it's not easy to get hold of stuff once it's in there. After you've layered up the bin with paper napkins, charging cables, travel sweets, bags, pens, wipes, coins and more, getting to the sugar sachet that you know is in there somewhere is a frustratingly drawn-out affair.

The solution is to get hold of some small containers that fit nicely into the storage area that you're aiming to compartmentalise and slide as many as you can into the space. Now all related items can be packed together, and they'll be found much more easily when you need them.

The other benefit of using these liners or dividers is that they're far easier to clean than the storage space that they sit in. You can simply whip them out, take them indoors and wash them with the dishes, ready for the next trip.

Coin tubes

Keeping loose change in the car is handy, until you need it at a toll booth or car park and suddenly you can't find any of it. There's some in the glovebox, centre console and door pockets, and definitely in the jacket that you left in the boot.

A good tip is to keep a selection of loose change in the car. A simple hack that kills two birds with one stone is to get a small cylindrical container, like a Berocca tube, chewing gum tub or sweet container, and use that for coins. It'll ensure that you always have cash in the car and that you know where it is, but it'll also disguise the money from any opportunistic thief should you leave a car door or window open.

TOOL HACK #6

Use an adjustable spanner as a screwdriver

When screws are tucked into hard-to-reach parts of your car, grip a screwdriver bit at right angles in an adjustable spanner to help get access. Or, even better, use a bit with a ratchet wrench – the hex shaft of most screwdriver bits fits directly into a ¼-inch socket.

CHAPTER 6

CLEANING HACKS

There are two events that make you feel super-positive about your car: one is when you get the phone call to confirm that it's passed its MOT, and the second is when it's been given a proper clean. The hacks in this chapter cover various elements of car care, from keeping it tidy inside and out, through time- and money-saving cleaning tips, all the way to advice for perfectionists who will never be satisfied until their car looks better than when it came off the production line.

Use olive oil to polish interior trim

The part of the car that the owner interacts with, sees and connects with more often than any other is the interior, specifically the dashboard and centre console. Therefore, if that's looking in mint condition it'll ramp up the feel-good factor, and a happy driver is a considerate driver so everyone's a winner. And the best way to hack a beautiful interior is with some olive oil straight from the kitchen.

Olive oil can be used for everything from loosening ear wax to conditioning hair – apparently some people even cook with it – but it's perhaps less well known that it gives the sort of plastics that are used inside cars a good shine and protect. Use sparingly on a clean, dry cloth and wipe it over the surface to achieve this.

Olive oil can also be used to smarten up tired-looking leather seats by applying less sparingly and leaving for half an hour before wiping off. Please note, however, that you shouldn't use olive oil on the gear knob or steering wheel, as it'll create a slippery surface that may be dangerous when driving.

In addition, be sensible about what type of olive oil you use for this hack – there's no point in emptying a full bottle of first-press extra virgin oil to smarten up your dash, because it's likely to be cheaper to get dedicated interior trim cleaner from the auto spares store. But everyday olive oil is ideal for this task, and it doesn't take a lot of it to work wonders with the plastics.

Use a blade to clean your windscreen

The car windscreen is a veritable magnet for all types of muck, and being right in the line of sight makes it virtually the first indicator of a filthy car. Washing the screen is fine, and using specialised glass cleaner is better – or even the classic window-cleaning hack of neat white vinegar and scrunched-up newspaper. But there will usually be some ultra-stubborn marks like tree sap and kamikaze bugs that are hard to shift, and often you don't see them until you've tidied away your cleaning gear and embarked on another journey.

The car hacker's solution is to have a razor blade or Stanley knife blade to hand in the car. These are brilliant at getting marks off glass, and will do no damage to the surface as long as you use them properly (and no damage to yourself as long as you use them carefully).

Holding the blade at around a 45° angle to the windscreen, carefully scrape the mark away. If you have the patience, all the glass in the car can be cleaned this way, but that's for the real enthusiast.

 Coffee filter dusters

When reading the instructions on cleaning products, they always advise the use of a dry, lint-free cloth to dust the area in advance of cleaning. The good news for car hackers is that you are likely to have a pack of these in your kitchen or workplace in the form of coffee filters.

Simply use a coffee filter just like a duster, and the fluff and dirt will cling to it until it's ready for disposal. Store a few in the car and they'll always be to hand too. Just remember not to use them to brew a cuppa afterwards.

TOOL HACK #7

Make a breaker bar with a length of pipe

Known as a 'cheater bar' by some, a wrench with a length of pipe over the handle acts as a perfect breaker bar for those tough nuts to crack.

DIY air fresheners

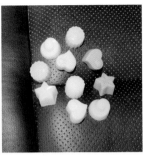

If your house smells good, it feels clean and fresh. The same goes for your car, and that feeling helps to arouse positive emotions. There are ways to get your car smelling nice without continually forking out on air fresheners and their refills or replacements. The first thing to do is regularly clear your car interior of any rubbish and clean any spills as soon as you can. But you can go a little further with the aid of a hack or two.

The herbal tea bag hack from Chapter 1 is a good start – dehumidifying gets rid of any musty odours and adds freshness. There are other dehumidifying tricks in that chapter, but if you want to add the pleasing aroma that an air freshener brings then try this. Get hold of an old jam jar or equivalent, and puncture the lid several times with a hammer and a pointed chisel or small crosshead screwdriver. Then put some perfumed wax (similar to above) in there and leave it in an unused cupholder in your car. As the car heats up in the sun, the wax will melt, releasing the aroma.

If you're more of an essential oils person, or it's a time of year when the sun isn't quite as reliable, rather than perfumed wax in a jar you can try this hack. Take a wooden clothes peg and drop some essential oils on the end. Clip the peg on to one of the car's central air vents, and when the fan is turned on the scent will be circulated around the cabin.

Cat litter moisture removal

Cat litter has two very distinct uses when it comes to driving, although as this chapter is concerned with cleaning and freshening the car, the focus of this hack is using cat litter to make the car more habitable.

I've spoken elsewhere in this chapter and beyond about the benefits of dehumidifying to ensure the car interior is unpolluted by stale, dank odours. Cat litter is ideal for imbibing moisture in small areas like a car, so fill a couple of muslin bags or even a pair of big socks with litter and place one on the top of the dashboard and one on the rear parcel shelf at night. The interior should remain immune to the condensation that results during temperature changes.

And that second kitty litter hack? Well, it also makes a very handy grit for when the car has its wheels stuck in snow. Pour a sizeable amount of the stuff in front of the driven wheels of your car and give it another try.

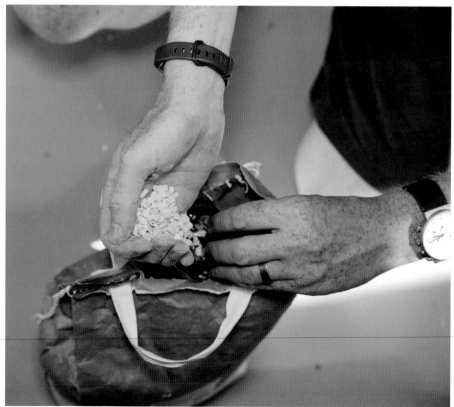

Use plastic breakfast cereal boxes as waste bins

Here's a great travel hack that is designed to ensure that the interior of your car remains tidy, clean and free from the lingering smells associated with litter. You know the type of thing – discarded cheese-flavoured crisp packets, old fast-food bags and wrappers, cans of fizzy pop … it's too easy to toss these whiffy food and drink carcasses into a footwell, a door pocket or on to the back seat and worry about them later. 'Later' ends up being several months later, as bits of this stinking detritus make their way into otherwise inaccessible areas between seats and the transmission tunnel, behind seat cushions and under the floor mats.

What you need is a waste bin – in fact, why don't all cars come with these as standard? But even small bins are too large for a car – and where to store it? The answer comes in the form of this hack, which involves a useful piece of kitchenware: a plastic cereal tub. These are narrow, tall, capacious and, as well as having a full-width lid for lining and emptying, they have a flip-top opening that makes access simple. This is perfect for its renewed car-trash purpose: the opening is easily accessible but it shouldn't spill its contents if it gets upturned while on the move.

A small pedal-bin liner should be ideal to repurpose the cereal box into a workable bin with wet waste capabilities. Line the tub as you would a regular bin, folding the excess over the top lip, then seal the lid on to secure the bin bag in place. Now put any travel litter in via the hinged access in the lid. This travel bin can be stored in the passenger footwell, and if you want to avoid it tipping over, a large elastic band or small bungee cord can be strapped round it and attached to the centre console's curry hook. If your car has no hook, you'll find out how to add one elsewhere in this book.

Make your own
grit guard bucket

Some see washing the car as a bit of a chore and others
as a chance to personally care for their vehicle, letting the
perfectionist in them out to play for a few hours, in order to make
their pride and joy look as good as it can possibly be. There are
also financial and preservational reasons to do this, some of
which are outlined elsewhere in this book, but the primary reason
should be that if your car looks good you feel good about it.
Washing the car yourself only multiplies that feeling.

There are some good pieces of advice about washing a car
that it would take too long to go into here. One major one
that's gained traction in recent times, filtering through from car
detailing and the show and shine scene, is using two buckets
(one for washing, one for rinsing the sponge/wash mitt) to avoid
contamination of the cleaning solution, and there's also the grit
guard bucket.

Rather than consign your old traditional bucket to landfill,
however, the anti-grit element can be hacked into it. Get hold
of an old plastic colander that's around the size of your bucket
(if you don't have one lying around, you can pick one up from

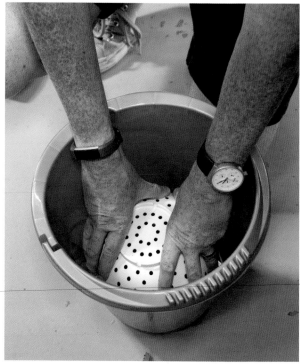

a charity shop or eBay for pennies). Shave the handles off with a craft knife or multi-tool so that it wedges in the bottom of the bucket. Now insert it upside-down in the bucket so that it looks like a dome in the bottom. That should do an excellent job of keeping the grit that you're cleaning off your car from staying on your wash mitt and scratching the paintwork.

If you search online, you'll find some even more ingenious self-made grit guards, so this hack can be taken as far as you want to go with it.

TOOL HACK #8

Use a pipe wrench to remove rounded bolts

Old and frequently used bolts can end up with rounded heads through wear and tear, which makes them a nightmare to remove. However, you can loosen them using a pipe wrench – its jaws dig in and grab the head so you can remove the bolt.

Use toothpaste to make headlights minty fresh

This hack is designed to solve a problem that affects some cars much worse than others: cloudy headlights. On some older cars this manifests itself as a milkiness or yellowing that affects the headlight plastic, and it can be so pronounced that the headlight beam is drastically reduced.

There are headlight cleaning kits out there, and this tip involves embarking on the same process as these kits outline – but rather than using costly cleaner, toothpaste comes to the rescue. Toothpaste's abrasive qualities make it ideal for gently polishing away the cloudiness, leaving clear plastic. Here's how to do it.

1. Clean the headlights to get rid of dirt and road film, or perform the toothpaste hack after giving the car a wash.

2. Get a tube of toothpaste. Any will do because they all contain an abrasive of some sort. Whitening toothpaste with soda would be ideal, but avoid anything containing microplastics.

3. Mask up the area around the headlights with masking tape and newspaper.

4. Smear some toothpaste on the first headlight with a dampened microfibre cloth. Gently rub a small area in a circular motion then build up pressure and keep rubbing until the headlight appears clearer.

5. Move on to the next area of the headlight, adding more toothpaste to the damp cloth as required. Note: if elbow grease and the cloth aren't cutting through the fog, try using a drill with a polishing mop attachment.

6. Go over the whole headlight again with an increased ratio of water to toothpaste, and then do the same again with even less toothpaste.

7. Once the headlight is looking crystal clear, wipe off the residue, rinse with clean water and dry with a clean microfibre cloth.

Baking soda for smelly upholstery

Upholstery can get a bit smelly over time, especially if something's been spilt on it, or the seats have been used to ferry bin bags; and let's not forget the ever-present problem of travel sickness. Wiping things clean deals with the immediate issue, but days and weeks later the odour can reappear.

The way to solve this problem is with some baking soda. Baking soda's household cleaning properties are so well known that it barely qualifies as a hack. However, its car upholstery-freshening abilities may not be as notable, so here goes. Sprinkle some on the affected area just like retro carpet-cleaning powder Shake n' Vac. If it's come out in lumps, spread it around the upholstery with a clean brush.

Leave it for half an hour or so, then get the vacuum cleaner out and vacuum up the baking soda. The upholstery should now have had its stink eliminated.

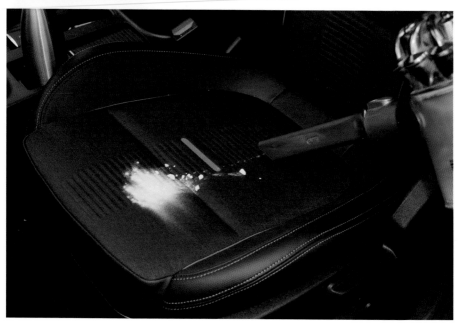

 ## Screwdriver cloth to replace Q-tips

When cleaning the car outside and in, there are little niches, corners and recesses that are incredibly difficult to reach – and these hard-to-reach areas are the places that are highly likely to get dirt and dust ingress. Embossed badges, panel joins and vents are just some of the areas that can be missed out by a run-over with the duster or cleaning cloth.

The answer to these problems used to be the alcohol-soaked cotton bud or Q-tip, but in recent times these plastic items have been discouraged from use in part because of their role in ocean pollution. Instead, try this hack. A microfibre cloth wrapped around the end of a small screwdriver is ideal for getting into those small or hard-to-reach areas, and can be reused again and again.

CHAPTER 7

MODERNISING HACKS

This chapter is for owners of older cars, and for the purposes of these hacks the definition of 'older' is approximately pre-21st-century vehicles (although some less well-specified or optioned newer cars can take advantage of these hacks too). There's an outline of what recent tech to add and how to add it, as well as some tips for adding cosmetic enhancements to give an older car some of those trendy touches. The sky's the limit when it comes to cosmetic enhancements, but this chapter concentrates on subtle touches rather than serious visual modifications – the latter are left to specialist magazines and websites.

The BMW MINI of 2001, and its subsequent models, is often credited with popularising the concept of factory personalisation, be that exterior stripes and contrast-coloured roof panels, or funky interiors with trim panels painted in the exterior body colour or given a metal-look or carbon-fibre-effect finish. Buyers could spec their car from a bewildering array of options, allowing them to express their individuality despite owning a mass-produced car, and this trend has become prevalent in the whole small car and mini-SUV sectors.

Owners of older cars or vehicles that didn't present the opportunity to personalise the interior need not despair at their plain and commonplace transport. Prising off a piece of trim and giving it a unique twist is simpler than you think.

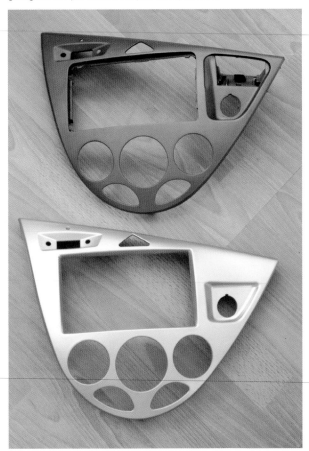

Popular pieces of trim to brighten up are the passenger airbag cover, air vent surrounds, centre console trim and instrument surround. Remove the pieces of trim as per the instructions that you'll find in the car model's Haynes Manual or online – generally it'll be no more than using a plastic pry, being careful not to break any lugs or delicate, slim areas of the trim.

Scuff up the surface with a Scotch pad, remove any dust with an alcohol wipe, mask up the clips to ensure they'll refit correctly and apply a coat or two of plastic primer. Wait until the primer has completely dried and now it's ready for painting – you can either colour match it to the car's exterior or try something like a gloss black or aluminium spray paint. Three coats should be enough, then for a shiny, durable finish apply a coat of lacquer once the paint is dry.

Reattach to its original position, and your car now has a personalised interior. If you want to spend a bit more to take it to the next level, why not try a water transfer print on the trim pieces? There are numerous companies that will do this now, and many will let you post the part to them and will post it back once it's ready. Water transfer printing, or hydrographics, puts a coating on your plastics that gives a carbon fibre effect or various other patterns.

Mirror/exterior trim wrap

A stylistic touch that has gained popularity over the last couple of decades is contrasting mirror caps. It was probably popularised by Audi, whose chrome or aluminium-look mirror caps became a signature of high-performance models in its range. Many manufacturers now offer mirror caps in contrasting colours to the bodywork as a personalisation option.

This is something that can be added to any car quite simply, using a small amount of vehicle wrap. This material can be bought in small quantities and in a variety of colours and finishes from various outlets or via eBay. The benefit of using wrap over respraying is that you gain an extra level of protection for these precariously positioned parts of the car, and the wrap can be removed if you change your mind or want to return the car to standard for selling on. Some wrap will come with an application kit, but if not you can do it yourself with a spray bottle, some washing-up liquid, a credit card, a Stanley knife and a hairdryer. First remove your door mirror cap – some will pop off if you get your fingernails underneath the edges, but more stubborn ones will require a plastic pry bar.

Make up a solution of water and washing-up liquid in the spray bottle and give a squirt to the mirror cap. Cut a piece of the vehicle wrap that's larger than the area to be covered, and remove the backing from the wrap. Using the hairdryer to heat the wrap, stretch it on to the mirror cap and use the credit card to smooth it on and remove any air bubbles (you may need another pair of hands here). The washing-up liquid will ensure that you can move the wrap so it's positioned correctly but will allow it to stick once dry. Once you're happy with the application, trim the edges with the knife. Now do the other one, and enjoy your Audified car!

Window tinting

Looking at some of the disastrous attempts at rear window tinting out there, it may be tempting to leave this to the experts, but with a bit of confidence and patience there's no reason why this can't be done at home. Window tinting has the practical benefits of providing a bit of privacy and (in many cases) also providing UV protection from the sun's rays, and certainly lessening the irritation of a low sun. It also gives a modern, classy look to a car – if done properly.

As every car is different, there's not enough space in this book to give detailed step-by-step instructions, and that's not the purpose of this chapter. However, here are some handy tips when embarking on the application of window tinting film.

Make sure that you buy good-quality window film from a respectable source, and that it's the right level of tint for the look you desire. Get hold of an application kit, if there isn't one supplied with the film. Make up a spray bottle of water and washing-up liquid to assist with the application – a light mist on the window will allow the film to be adjusted and assist with removing air pockets.

Remove the little black dots that you find around the edge of rear side windows that are not part of the door. These are purely aesthetic and are designed to blend the black window frit into the clear glass. However, they sit proud of the glass and that means the tinted film won't stick well to the window edges, leaving a noticeable rim to your freshly tinted glass. You can try removing them with a window scraper, or sharp blade, along with some elbow grease and patience, but a more effective method is to paint over them with black enamel paint spray, masking up either side of the dot matrix carefully.

Please note that tinting the front windscreen is illegal, and there are tight regulations about how dark the front side windows can go. Stick with the rear side windows and tailgate glass and you should be fine.

Headlight updates

One way to give a car a modern look is to upgrade the headlights, and to add daytime running lights (DRLs), which are those 'always-on', often LED, lights that have been fitted to most new cars and vans since 2011 to make them more visible in daytime conditions.

You can add DRLs to any car, or even convert foglights to DRLs. The simplest way to do this is to get hold of a DRL conversion kit; no shortcuts should be taken when it comes to safety equipment like lights. Look for an approval mark on the kit that includes the letters 'RL'. If you're fitting daytime running lights they should be installed so that they come on with the engine and go off when headlights are turned on.

Headlight upgrades are many and varied. A great tip for modernising your car is to look for upgraded headlight units, such as projector headlights or xenon units, that were an option or on upmarket variants of your car's model year. You can pick these up second-hand from scrap yards and eBay, and they are often 'plug-and-play'; in other words, you can simply replace your old headlight units with the new ones and they will work straight away.

A great simple hack is to darken the headlight within the unit. Once again, headlight units vary from manufacturer to manufacturer, and we're talking about multi-lamp units here that appear on cars from around the early 2000s onwards, not single headlamps. You don't paint over headlamp reflectors. This is how it's done.

EQUIPMENT

Gloss or satin black spray paint

Heat gun

All-weather silicone sealant

Craft knife

Masking tape

Newspaper

Whatever tools are required to remove the headlamp unit.

3. Mask up the bulbs/reflectors

Using masking tape and newspaper (if required – often just a lot of masking tape will suffice), cover up all the areas that must not be painted, for example bulbs, reflectors or projector lenses. In some units, all the light architecture can be removed, leaving the aesthetic brightwork that can be painted.

4. Paint

Rough up the area to be painted with a Scotch pad and wipe away any dust with an alcohol-soaked lint-free cloth. Shake the spray paint can as per the instructions, and test the spray on a piece of newspaper. Make one pass trying to cover all the little nooks and crevices and let it dry. Apply another coat and, if happy, leave it to dry for a few hours.

5. Reassembly

Clean up the inside of the glass with a microfibre cloth, and reassemble the innards of the headlamp assembly. Using the silicone sealant, reseal the headlamp unit and let it set. Then put it back on the car and repeat for the other unit.

1. Remove the headlight units

This step is unique to your car, and its Haynes Manual may instruct you on the removal of headlight modules.

2. Remove the glass lens

The headlight unit will have a seal around the lens. Using the heat gun, warm the sealant until it is soft enough to remove the glass lens. Carefully pry off the lens slowly so as not to damage it.

Many new cars offer Wi-Fi internet connectivity as standard, and this proves particularly useful when travelling with multiple people and all their tablets, laptops, phones, handheld gaming and more. Although many new-car buyers will think that this is brought to them by magic pixies and moondust, the secret is that in-car Wi-Fi is nothing more than a cellular antenna and SIM card, just like the one in your smartphone. And you can get hold of this for any car for the cost of a few takeaway coffees per month.

Most of the big-name mobile service providers offer a mobile broadband router (known by various other names, including Mi-Fi and Wi-Fi dongle) of one kind or another. Various plans can be selected that suit many different usages, and some even offer a pay-and-go type SIM, which is ideal for those only looking to use it on occasional vacations and long journeys.

The hacky element to this is to find an area of the car to get a good signal while remaining inconspicuous – the last thing the car hacker wants is to have an obvious accessory sliding around the dashboard in plain sight. A high vantage point is preferable for a mobile data signal, so does your car have one of those sunglasses holders next to the map-reading lights? Or a flap in the sun visor to store paperwork? If so, it might keep your dongle hidden. If not, try under the dash top – removing the glovebox lid and internals using instructions from a Haynes Manual or equivalent should give you a bit of space to attach the dongle. A double-sided sticky pad will attach it to the underside of the dash, or you could use zip ties to strap it to a dash support bar. Charging can be done by splicing into a power source behind the dash with a USB socket.

Get in-car performance data readouts

A surprising hallmark of modern cars is that drivers are given vast swathes of data about how the car is running. Many moons ago this would have taken the form of a few gauges, giving fluid levels and temperatures, plus pressure and turbo boost dials adding extra info where required. But customers are now demanding information on an unprecedented scale. This probably began with the popularity of hybrid and electric cars, which bombard the driver with efficiency information, often in pleasingly novel graphic forms.

In many new family cars, the screen in the centre console, once the home to audio controls alone, divulges data gleaned from the car's ECUs that includes everything from G-forces through live fuel consumption to engine diagnostics. And you don't just find this on performance cars – the humble Mk 7 Volkswagen Golf is just one modern family car that provides this bewildering array of onboard intelligence.

However, any car with an ECU and an onboard diagnostics (or OBD II) port can get hold of this information and have it transmitted to some aftermarket in-car entertainment screens or – even car hackier – to a smartphone.

A Bluetooth OBD II reader, available from many auto parts retailers, plugs into the OBD II port of the car, and an Android or iOS app will translate what is beamed to your phone into graphical data. Mount your smartphone to the dashboard with any of the hacks found elsewhere in this book, and you can now get performance and diagnostic information just like you'll find in more modern machinery.

Smartphone dashcam

Dashcams have divided drivers from those who marvel at their safety and insurance benefits to those who despise the assistance they give to highway vigilantes and YouTubers filming reckless road-based antics. Whatever way you slice it, dashcams have changed driving for good. A good dashcam requires a reasonable outlay, but there's a hack that will provide a decent onboard camera for a fraction of the cost. It involves an old smartphone, but if you're the type to trade in or sell on your old phones you could always buy a cheap older phone on eBay for this purpose. Why wouldn't you use your current smartphone? It's best to permanently attach the phone dashcam to the car and use it for this sole purpose, so it's not interrupted by calls, notifications and music.

Rather than hacking a mount for the smartphone, for this purpose I'd recommend buying a strong windscreen suction mount to ensure the best placement, which should be central and positioned so that the camera captures as much of the road ahead as possible. It'll also require a permanent power source while driving, so to avoid a dangling cable get hold of a power cord compatible with the phone and route it around the roof lining, A-pillar trim and the back of the dashboard by prising up the edge of each and pressing the cable behind, while you can use good old binder clips to route it down the dashboard to the 12v outlet.

The phone will require software, known as black box apps. There are several available for iOS and Android – search 'black box car' in the App Store or Play Store – and they all provide these features: continuous camera recording, location tracking and data collection, including estimated speed, date and time. In order to create the maximum storage space on the smartphone, delete all unnecessary apps, music, video and photos, and adjust the video quality settings in the black box app to allow decent recording quality while reducing file size.

Add aux input to stereo

The most reliable method of connecting an MP3 player or smartphone to an in-car audio setup is via a wired connection, yet older cars' hi-fis don't feature this. One way around this is to replace the car stereo with a newer one, but an alternative is to add an auxiliary input to the existing unit.

This is a fairly simple operation. Pull the stereo out of the dashboard and have a look at the input connectors at the rear. Most will have RCA jacks, which are the red and white female connectors that are used to connect external audio sources, just like in home music devices. Now get hold of a 2-metre male RCA to male 3.5mm adapter cable. Plug the RCA ends into the stereo and route the 3.5mm end down the back of the dashboard (you may need another pair of hands here). You can route it into the glovebox or even hack the dashboard's ashtray to route it through the back, and use the ashtray as a phone mount.

Add USB sockets in rear

USB sockets are not just handy but near-essential in a family car at this point in the 21st century – they are much more compact than 12v sockets and will charge any portable electrical device around. No wonder that 12v-to-USB adapters have emerged as a popular aftermarket accessory, but they have issues – mainly that they take up an otherwise useful 12v socket and can also work their way loose when multiple USB cables are hanging out of them into the cabin, limiting their real-world usefulness.

Better to have USB sockets built into the rear of the car's centre console so that they can be accessed from the rear seats. This can easily be done by adding a twin USB panel into the appropriate place – an ideal location is adjacent to the existing rear 12v socket, if the car has one.

There are kits available for this, and they only cost a few pounds. They should contain all the wiring you need, a 12v to 5v converter and the USB panel. The kit's instructions should be followed as each will vary based on the kit you buy and what your car looks like, but the basics are that the converter is connected between an existing 12v cable and the USB sockets. Drill an area for the sockets to go, remembering that there will need to be a large gap behind for all the connections and cables, and slot it all in.

OEM+ upgrades

Upgrading a car so that it has the equipment and specification of a higher-end model in the range is known as OEM+ modifying, referring to original equipment manufacturer plus. Basically, it's nicking bits off the posh version of your car.

Going into detail about every upgrade that could be performed on every car range would take several books, but here are some suggestions.

Alloy wheels – there are some exceptions, but the wheels from the sportiest models in the range should have the same bolt pattern as every other model in that range – but remember to check the offset. Buying second-hand and refurbishing the wheels yourself is the canny hack here.

Steering wheels – upgrade to a leather steering wheel by nabbing one from a luxury model. It should be a straight swap.

Seats and interior trim – second-hand leather sports seats and the matching door trims from top-spec models can be picked up from breakers' yards and eBay. Be aware – replacing seats and trim is a laborious job!

Fog lights – adding fog lights from an up-range model can be simpler than you think, because the wiring is often right there in the loom!

Body-coloured mirror caps/ exterior trim – this is as simple as searching on eBay.

Cruise control – requires a module as well as a switch, but this upgrade is not outside the limits of the home mechanic.

Intermittent wiper – upgrading these can simply be a case of getting hold of a second-hand intermittent wiper column stalk from eBay.

Brake upgrades – Fitting bigger discs can be a case of swapping in the calipers from a sporty version of your car, then ordering the correct discs and pads for that model.

Trip computers – these can be added very simply to many makes and models of car, but be aware that they will also require the addition of temperature sensors, throttle sensors, etc.

TOOL HACK #9

Extend a cordless driver with a four-way screwdriver

To get into the darkest recesses for screwing and unscrewing, you can extend the reach of your electric screwdriver by pulling the shaft out of a four-way screwdriver and clamping it into the chuck.

Add interior LEDs

Easy-to-fit LED interior lighting strips are highly effective at making the inside of your car easier to use at night and add a modern feel to the interior. LEDs draw little current, meaning they can be spliced into existing wiring without worrying about overload.

LED strips are available from various sources, from auto parts suppliers to eBay. The best place to start is by putting a pair under the dash on the driver and passenger sides, and once you're confident with that, the sky's the limit...

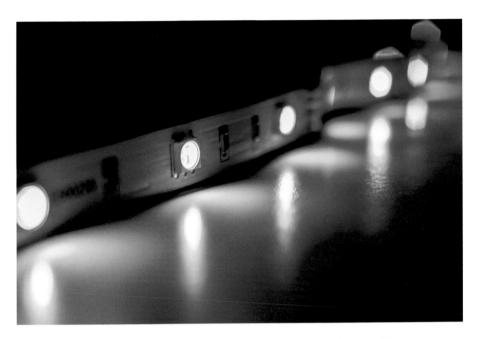

Choose a fuse

Find your car's fusebox and decide how you want the LEDs to come on. A good option is to use the fuse for the ignition, so that the lights come on when the key is turned and switch off when the car is shut down.

Splice into the wire

Now splice into the wire for the ignition fuse. Locate the live side of the fuse using a multimeter. Turn the ignition key on, earth the black probe and touch the red probe on one side of the fuse while it's still in the fusebox. If no voltage is detected, then test the other side. Splice into the side without the voltage using a connector.

Find an earthing point

Look under the dashboard to find an earthing point. You should find an area where electrical parts are bolted to the firewall, so you can loosen one of those and add a wire with a connector.

Add another fuse

Insert an inline fuse to the wire that you have spliced into the ignition fuse, for added safety.

Connect the LED strip

Now join the wire from the fusebox and the earthed wire to the first of two LED strips. While doing this, add another set of wires to this connection for the second LED strip. Once this connection is made, route the wires under the dash to the other side of the car.

Stick on the strips

Most LED strips like this are self-adhesive, so it should be a case of removing the protective backing and sticking the strip to the relevant area under the dashboard. Be sure to clean that area thoroughly with alcohol and leave to dry before offering up the strip.

Conceal any loose wiring

You should be able to tuck any straggling wires under the dashboard, but have some duct tape and zip ties handy to strap them out of harm's way.

Author Acknowledgements

Thanks to Rebecca and Archie for the support and putting up
with me while I wrote this book.

Craig Stewart has asserted his right to be identified
as the author of this work.

First published in November 2019

British Library Cataloguing in Publication Data
A catalogue record for this book is available
from the British Library.

ISBN 978 1 78521 651 0

Library of Congress catalog card no. 2019942919

Published by Haynes Publishing,
Sparkford, Yeovil, Somerset BA22 7JJ, UK
Tel: 01963 440635
Int. tel: +44 1963 440635
Website: www.haynes.com

Haynes North America Inc.
859 Lawrence Drive, Newbury Park,
California 91320, USA

Printed and bound in Malaysia

While every effort is taken to ensure the accuracy of the
information given in this book, no liability can be accepted
by the author or publishers for any loss, damage or injury
caused by errors in, or omissions from the information given.

CREDITS

Cover
Jud Webb

Page design:
Simon Larkin

Photography:
Ali Jennings
Jeff Meyer
Craig Stewart
Shutterstock